# Stings, Wings and Crawly Things.

by

David Wright.

Print ISBN 978-1-9195514-5-6



Published by

Llyfrau Cambria Books, Wales, United Kingdom.

Cambria Books and Cambria Stories are imprints of

Cambria Publishing Ltd.

Discover our other books at: www.cambriabooks.co.uk

# Acknowledgements.

This publication is the third in a series dealing with the relationship(s) between organisms and their natural environment. Earlier works were entitled "*Natural Toxins. The Good, The Bad and The Deadly*" and "*Fungi. Animal, vegetable, or what?*". This book focuses on the often-hidden world of invertebrates that share our planet in numbers that far exceed other animals.

I am again greatly indebted to Sarah Hannis for her help with illustrations and her enthusiastic advice concerning some of the six-legged, winged individuals.

Above all I am grateful to my wife Susan for her constant support and for coming up with the title.

# CONTENTS

# Introduction.

## Phylum Arthropoda.

The animals discussed in this section are all members of the phylum Arthropoda, the largest group of animals on earth, comprising more species than all the other phyla combined. Despite uncertainties in classification arising from shifts in taxonomic affiliations over time, the Arthropoda is currently divided into several major groups or subphyla depending on their body segmentation, the number and type of appendages and habitat. Although these subdivisions have been subject to a degree of historical variation that persists to this day, the four major living subphyla are generally recognised as Chelicerata, Hexapoda, Crustacea, and Myriapoda, The subphylum Trilobitomorpha is also recognised as an Arthropod subphylum (TABLE 1.), although this group re now extinct and members are only recognised from the fossil record.

## TABLE 1. Major Subphyla and classes of the phylum Arthropoda.

| Subphylum | Class | Common name |
|---|---|---|
| Chelicerata | Arachnida, | Spiders, Mites, Ticks, Scorpions |
| | Merostomata (Xiphosura), | Horseshoe Crabs, |
| | Pycnogonida | Sea Spiders |
| Hexapoda | Insecta | Winged: Pterygota Wingless:Apterygota (Entognatha[1] ) |
| Crustacea | Brachiopoda | Lamp Shells |
| | Maxillopoda | Copepods, Barnacles |
| | Malacostraca | Crabs, Crayfish, Lobsters |
| Myriapoda | Chilopoda | Centipedes, Millipedes, Pauropoda, Symphyta |
| Trilobitomorpha (now extinct) | Trilobita | Trilobites |

In the following sections the major subphyla and their respective classes are summarised with emphasis on specific taxonomic groups that have made major contributions to science and human culture.

[1] Entognattha are a class of wingless Arthropods originally included under the Apterygota as members of the now obsolete order Thysanura. It is now considered that Entognatha and Apterygota (TABLE 2) are paraphyletic, i.e. they share a common ancestor.

# Subphylum Chelicerata.

The subphylum Chelicerata takes its name from the ancient Greek word for claw which aptly describes several of the multiple appendages that these animals possess. Chelicerates are noted for their segmented bodies consisting primarily of an anterior cephalothorax and a posterior abdomen. In addition to limbs use for walking (or in rare cases swimming) other appendages are modified for the capture of prey or for sexual activity.

Chelicerae, often referred to as mouthparts, may take the form of articulated fangs or pincers. In many chelicerates, including most spiders, these fangs may be associated with venom glands and are used to inject venom into prey or as a defence against potential predators. Palaeontologists have discovered fossilized examples of Chelicerates complete with pincer-like appendages dating back to the early Ordovician era which lasted from nearly 500 to 440 million years ago. The earliest example, *Megachelicerax cousteau*, (FIGURE 1.), which is named after the famous oceanographer Jacques Cousteau, is estimated to be about 500 million years old.

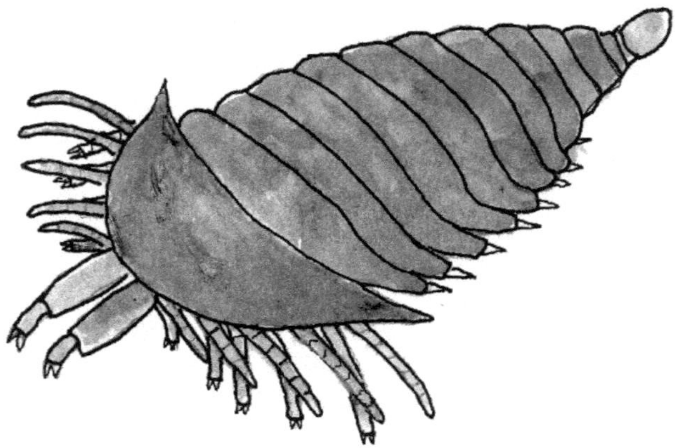

FIGURE 1. ***Megachelicerax cousteau,*** believed to be the earliest member of the Chelicerata, discovered as a fossil in a Utah, U.S.A. desert but of marine origin

Approximately 80,000 living species of Chelicerates have been described, making them the next most abundant group behind the insects (Hexapoda). They include more than 50,000 species of spider. Although more than 30,000 ticks and mites, known as Acari, have also been identified, their small size and elusive habitat make them difficult to find, and it has estimated that the total number of species in the Chelicerata subphylum probably exceeds 120,000.

# Class Arachnida

## Spiders.

Spiders possess two major divisions of the body: the cephalothorax and the abdomen. The cephalothorax has been described as a fusion of the head and the thorax wherein the 'head' and 'chest' regions have been combined into a single entity. However, there remain some differences in interpreting the segmentation of the head and thorax. The heads of most spiders have eight eyes, consisting of one pair of primary eyes and three pairs of secondary eyes.

Spiders have no antennae, no wings and possess four pairs of legs that grow from the cephalothorax together with a pair of palpi. Palpi are claw-like appendages that, in female spiders, are used for physical exploration of the environment. In male spiders the swollen palpi are modified as sexual organs that transfer packaged sperm in the form of spermathecae to the oviduct located beneath the abdomen of the female spider. This transference of sperm carries significant risk for the male spider who is in danger of being eaten by the much larger female during the act. Not the best honeymoon for the male of the species!

Another feature of a spider's abdomen is the spinneret, which is an organ responsible for producing the silk used for web-making. The web is, of course, inextricably linked to the spider's primary mode of feeding which mainly consists of ambushing and trapping small insects. Although a few spiders are active hunters, the entrapment of insect prey in webs is the main method of obtaining food for most spiders.

The chemical compounds that form the spider's web are a large group of scleroproteins which are elongated polypeptide chains consisting of approximately 3500 amino acids, notably glycine and alanine. The chemical components in these complex structures, known as spidroins, are configured in a manner that incorporates exceptional strength and flexibility. This remarkable combination is illustrated by the ability of these silk fibres to stretch by as much as 135% without breaking. It is, perhaps, no great surprise that these qualities have recently caught the attention of the biomedical industry interested in developing strong but flexible material for tissue regeneration and repair. Within the last fifteen years or so, spidroins more or less identical to spider's web silk fibres have been artificially manufactured using a genetically modified version of the marine bacterium *Rhodovulum sulfidophilum*. An attractive feature of this application of biomanufactured spidroins is the low incidence of immune response resulting from their use as regenerative agents for a variety of organisms. This reduces their tendency to be rejected as 'foreign bodies'. Another quite different projected application of manufactured spidroins is their use in the manufacture of bullet-proof tunics; another clear endorsement of their extraordinary tensile strength.

Larger spiders can handle bigger prey organisms, although their mouths and alimentary systems have little capacity for dealing with solid food. They therefore rely heavily on the secretion of enzymes to initiate the external digestion of prey through enzyme secretion. This enables food to be ingested in liquid form.

In terms of leg span the world's largest, or at least the broadest, spider is probably the Giant Huntsman *Heteropoda maxima* (Order Araneae; Family Sparassidae) (FIGURE 2) which can span as much as 30cm (12"). As its name suggests, it is a hunter rather than a web-builder. It mainly feeds on insects and occasionally small rodents but rarely bites humans unless unnecessarily provoked. Its bite is non-fatal to humans but can cause nausea.

FIGURE 2. The **Giant Huntsman spider** *Heteropoda maxima* is the world's largest spider in terms of leg span, which can be up to 30cm (12"). They are native to east Asia. *This specimen, reported in the Daily Express newspaper was photographed in Hull U.K. aboard a Chinese container ship.*

For sheer bulk the tarantula spiders probably hold the record with a recorded weight as much as 6oz (170grams). The largest tarantula specimens may reach at least 3" in body length with leg spans up to 10". Tarantulas belong to the family Theraphosidae (order Araneae; Infraorder Mygalomorphae) and comprise as many as 1,200 species belonging to 182 genera.

## Pet Tarantulas?

The Mexican Redknee Tarantula, as its name suggests, is native to Mexico, although the name Redknee has been applied to two large con-generic species, *Brachypelma smithi* and *B. hamerii*. *Brachypelma smithi* (FIGURE 3) was first described by British arachnologist Frederick Pickard-Cambridge in 1897. Despite their large size and intimidating appearance tarantulas are relatively mild in character and rarely bite humans unless provoked. Their venom is relatively weak and seldom produces a reaction more painful than a bee sting. Their docile demeanour coupled with gentle handling has even led to them being kept as pets. However, this is understandably a rare occurrence!

FIGURE 3. **Mexican RedKnee Tarantula** *Brachypelma smithi* has a total leg span of about 6".

## Carolina Wolf Spider.

By comparison with the tarantula, the largest Wolf Spider in North America is the Carolina Wolf Spider (*Hogra carolinensis*) (FIGURE 4) which has a body length of 1.4" and a leg span of 3-4".

FIGURE 4. **Carolina Wolf Spider** (*Hogra carolinensis*)

Although the Carolina Wolf Spider is not regarded as harmful to humans, speaking from personal experience they are guaranteed to grab your attention if they land on you.

## Funnel web spider.

Australian funnel web spiders are another group of stout legged robust bodied spiders belonging to the infraorder Mygalomorphae. Although they are not as big as tarantulas, funnel web spiders (FIGURE 5) are among the deadliest of all arachnids. Funnel web spiders belong to three genera of the family Atracida; Atrax. Hadronyche and Illawarra. All three genera are native to Australia.

FIGURE 5. **Australian Funnel-web Spider**. *Atrax robustus.* Among the most poisonous of all spiders. .

## Mites.

These tiny members of the Arachnida class, with a life span of about a month, are cohabitants of the human world that were first described by French zoologist Edouard Trouessart in 1897. The two most common species are dust mites of the Order Sarcoptiformes, Family Pyroglyphidae. These two species, the European Dust Mite *Dermatophagoides pteronyssinus* (FIGURE 6.) and the American Dust Mite, *D. farinae*, account for over 90% of this taxonomic group. They are to be found among human clothing and bedlinen although 'found' is probably not the most appropriate word considering that they are only 0.2-0.3 mm in length, comparable with a single grain of salt! Although they usually live on or near human skin, they do not possess biting mouthparts and can perhaps best be described as scavengers of dead skin cells. Although they don't bite, they are nevertheless irritants and can cause allergic reactions in some susceptible individuals.

FIGURE 6. **The Dust Mite** *Dermatophagoides pteronyssinus* is a nuisance species of the Class Arachnida

# Class Merostomata (Xiphusura).

## Horseshoe Crab.

The Merostomata, also known as Xiphusura are an ancient class of chelicerates with a fascinating history and a rich cultural heritage. The only living members of the Merostomata class belong to the genus *Limulus* which gives its name to the Limulidae family. The common name for this marine organism is the horseshoe crab. Despite the name 'crab' its genetic makeup places its taxonomy closer to spiders and scorpions than crabs – or any other crustacean.

The 'horseshoe' part of its name is a direct reference to the shape of its carapace (shell) which is not unlike a horse's hoof. The body parts of the horseshoe crab are divided into three sections; the prosoma (head), opisthosoma (abdomen) and telson (hindmost section or tail). The telson takes a variety of forms and functions in arthropods including those of a defensive stinger or as a tail fan capable of facilitating an escape reaction when required. In the horseshoe crab the telson takes the form of an elongated spike that apparently functions as a lever enabling it to change direction or flip over as required. In addition to appendages that are modified for feeding or mating behaviour, horseshoe crabs have four pairs of legs that are actually used for walking. Among other physical characteristics are no less than 10 eyes including a large pair of compound eyes that have contributed to their reputation for supernatural vision and insight in native North American folklore. This is in spite of the fact that their actual eyesight is not very good apart from hypersensitivity to light.

There are four living species of horseshoe crabs, three of which are native to southeast Asia. The only American species is *Limulus polyphemus*. They are slow-growing animals that take about 10 years to grow to adulthood with a normal lifespan of 20 years.

*Limulus polyphemus* (FIGURE 7.) is an interesting feature of the shallow water communities of the U.S. Eastern States New Jersey, Delaware and Maryland where they feed on marine and estuarine crustaceans and other invertebrates. Their elaborate seasonal mating behaviour involves the smaller male climbing on top of the larger female, often above the high-water mark. Mating culminates in a nesting and egg-laying ritual wherein batches of eggs are laid in shallow sandy nests close to the water's edge. This behaviour makes their eggs 'fair game' as food sources for a variety of shorebirds and other predators. Shore birds such as the Red Knot even time their annual southerly migration to the Chesapeake Bay to coincide with the seasonal (spring and summer) mass invasion by horseshoe crabs, thereby ensuring a reliable food source to nourish the migrating birds on their long migratory journey . They are also a target for marine biologists, including the author of this book, anxious to witness the seasonal mass migration and shoreline invasion by these extraordinary creatures.

Of the four known species of horseshoe crab only the Indo-Pacific horseshoe crab (*Tachypleus gigas*) is considered safe for human consumption, although regional conservation regulations may apply. In any event consumption remains a high risk enterprise. The Mangrove horseshoe crab (*Carcinoscorpius rotundicauda*) contains the deadly neurotoxin tetrodotoxin that causes fatal poisoning even when cooked. It, therefore, goes without saying that safe consumption requires correct species identification. Overall, it is probably safer to follow the lead of seabirds and limit consumption to the animals' eggs.

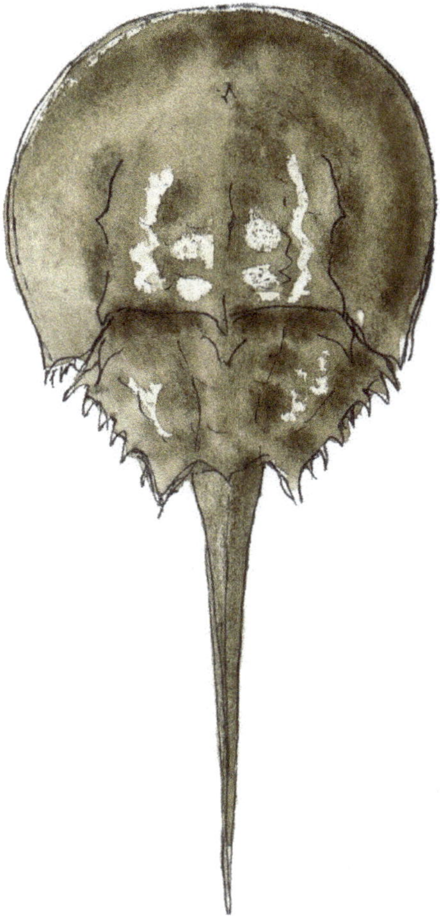

FIGURE 7. The **Horseshoe Crab** *Limulus polyphemus* is the sole North American species of horseshoe crab. It is a seasonal visitor to the east coast of the U.S.A. with a long folklore history.

Due in part to their ancient evolutionary history, their peculiar mating behaviour as well as their unique anatomy horseshoe crabs have been the subjects of a worldwide cultural heritage with many facets.

Among their most notable characteristics is their ancient heritage going back to a period 450 million years ago, before the advent of the dinosaurs. The unusual anatomy of both fossilised and living specimens has enabled us to track their fossil history, which tells us that their anatomy has remained remarkably constant throughout that era, thus earning them the accolade 'living fossils'.

Because of their robust but detailed exoskeleton, fossilised horse crabs (FIGURE 8) retain many of the detailed features that are found in living specimens.

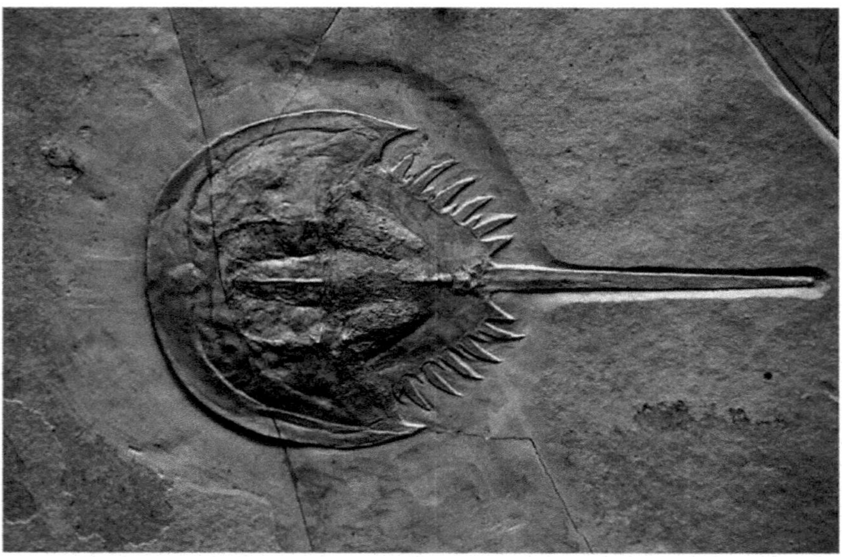

FIGURE 8. **Fossilised North American Horseshoe Crab.**
*Courtesy Florida Fish and Wildlife Conservation Commission.*

The tenacious grip of the male horseshoe crab on its female partner has been given a romantic bias in eastern Asian folklore which confers a 'husband and wife' identity to the coupling. In Malaysia, Thailand and Vietnam the inseparable clinging of the horseshoe crab lovers is invoked as an ideal; a metaphor for 'never let go' particularly following separation.

However, the legacy of the horseshoe crab is not all 'Love' and 'Romance'. The helmet-like shape of the horseshoe crab carapace is probably behind some of the more warlike symbols that have been associated with the horseshoe crab. In Japan the horseshoe crab shell has been used as a painted symbol of the battle of Dan-no-ura. This was a decisive naval battle of the Genpei War fought in 1185 in the Kanmon Straights between Honshü and Kyüshü. The defeat of the Taira clan by the Minamoto family led to the rise of the Samurai-led shogunate. Horseshoe crab shells are regarded as a memorial to the Minamoto dead, but also a celebration of Buddhist reincarnation of those who lost their lives in the battle. A more spiritual role for horseshoe crab shells is embodied in their use in some Chinese communities as a means of warding off evil spirits. Horseshoe Crab shells with painted tiger heads are hung above doors to keep away malevolent demonic influences,

In addition to their rich folklore history horseshoe crabs are now known to possess an important asset for the biomedical industry related to their copper-based blood. The characteristically blue blood[2] of horseshoe crab is known to contain the substance Limulus Amoebocyte Lysate (LAL). LAL

---

[2] The blue colour that characterises the blood of many invertebrates including the chelicerates is due to the presence of haemocyanin, a pigment that has a similar function as haemoglobin in the blood of vertebrates. The blood of arthropods is, therefore, usually referred to as haemolymph.

is a cell suspension that is highly sensitive to bacterial toxins. In the presence of these endotoxins LAL will coagulate; a response that signals the presence of bacteria. LAL therefore acts as an indicator of bacterial contamination that is widely used to test the sterility of surgical instruments. In the U.S. the Food and Drug Administration currently mandates the use of LAL to test all injected and intravenous drugs for gram negative bacteria.

# Class Pycnogonida.

## Sea Spiders.

Although these arthropods bear a strong resemblance to land-dwelling spiders, notably their possession of eight legs, they are placed in an entirely different class (Pycnogonida). than their terrestrial relatives (class Arachnida). They do, however, share the common name spider, albeit 'sea spider'. Within the Class Pycnogonida, sea spiders comprise a single living order, the Pantopoda containing about 1300 species divided among 11 families. They are widely distributed in most ocean systems throughout the world.

Although most sea spiders are less than 1 cm in diameter (FIGURE 9), some giant forms have leg spans that may exceed 50cm (1.5') (FIGURE 10). Their most striking characteristic is their exceptionally long legs relative to their bodies. Their ancestry dates back as far as the Cambria era of 539-487 million years ago although their evolutionary lineage relative to other chelicerates remains a topic of debate.

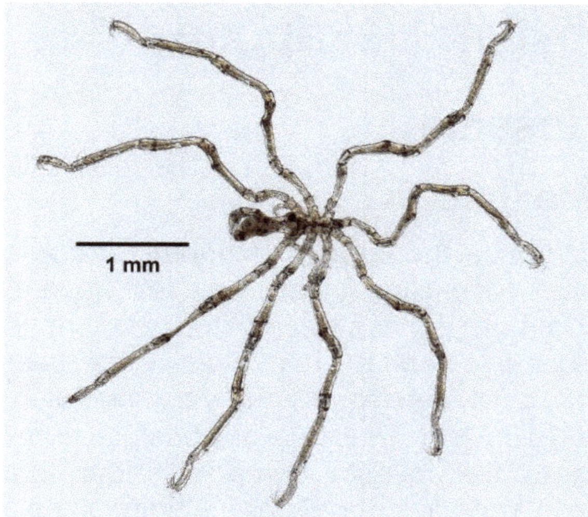


FIGURE 9. **Sea-spider.** *Callipallene brevirostris.* Long Island Sound, Connecticut. U.S.A. *Photograph by Eric Lazo-Wasem.*

FIGURE 10. **Pycnogonid Giant Sea Spiders** may exceed 50cm in diameter.

# Subphylum Hexapoda.

## Class Insecta.

### What Are Insects?

Insects comprise the largest taxonomic group of animals on earth. Over 1.2 million insect species have been formally identified, although it has been estimated that, in view of their often-concealed habitat and evasive behaviour the true number of insect species may be very much larger and could approach 10 million. In terms of absolute numbers, animals belonging to the Insecta comprise more than half of all worldwide eukaryote[3] organisms including animals, plants and fungi. Although some insects are capable of swimming, they are essentially terrestrial, air-breathing organisms. Other terrestrial arthropods include centipedes, millipedes, mites, scorpions, spiders and woodlice (TABLE 1).

For taxonomic purposes insects (class Insecta) are often divided into winged (Pterygota) and wingless (Apterygota) forms. In some publications Apterygota appears to be synonymous with Endognatha which has also been defined as a (wingless) class within the Hexapoda. Whether or not Apterygota and Endognatha are exactly equivalent there is general agreement that they are paraphyletic i.e. they share a common ancestor. Members of this class, however it is defined, are Collembola (springtails), Diplura (two-tails) and Protura (coneheads).

---

[3] A eukaryote is a cell or organism that possesses a clearly defined nucleus surrounded by a membrane that also contains hereditary material in the form of chromosomes. This differentiates them from prokaryotes, which are unicellular organisms with no nucleus (e.g. bacteria).

## TABLE 2. **Taxonomy of Winged Class Insecta.**

| | |
|---|---|
| Class **Insecta.** | |
| Subclass **Pterygota** (winged insects) | |
| Infraclass **Neoptera** | |
| **Endopterygota** (Holometabola) | **Exopterygota** (Hemimetabola) |
| Orders. | Orders. |
| Coleoptera (Beetles) | Anaplura (Sucking Lice) |
| Diptera (Flies, Mosquitos) | Dermoptera (Earwigs) |
| Hymenoptera (Bees, Ants, Wasps) | Dyctioptera (Cockroaches, Mantids) |
| Lepidoptera (Butterflies, Moths) | Ephimeroptera (Mayflies) |
| Mecoptera (Scorpion Flies) | Hemiptera (Bugs) |
| Neuroptera (Lacewings, Ant Lions) | Homoptera (Aphids) |
| Siphonoptera (Fleas) | Isoptera (Termites) |
| Tricoptera (Caddis Flies) | Malophaga (Chewing Lice) |
| | Odonata (Dragon Flies) |
| | Orthoptera (Crickets, Grasshoppers) |
| | Phasmida (Stick/Leaf Insects) |
| | Plecoptera (Stone Flies) |
| | Thysanoptera (Thrips) |

Further taxonomic subdivision of insects is shown in TABLE 3.

TABLE 3. **Taxonomy of Wing-less Arthropods that are often (but not universally) classed as insects.**

| Class **Insecta** |
|---|
| Subclass **Apterygota** (wingless 'insects') |
| **Orders.** |
| Archaeognatha (Bristletails) |
| Collembola (Springtails) |
| Diplura (Camboda Silverfish) |
| Protura (Coneheads) |

The primary anatomical characteristic of insects is their division into three distinct body segments: the head, thorax and abdomen. Compared with other terrestrial arthropods only insects are capable of flight, provided by one or two pairs of wings that emerge from the thorax. Insects differ from other arthropods in the possession of three pairs of legs, which are also thoracic appendages. These six legs are responsible for the name of the insect sub-phylum Hexapoda.

The anatomical division of the insect's body into three distinct segments probably accounts for the origination of the word Insecta which is derived from the Latin 'Insectum' meaning 'Cut Up'. This in turn has given rise to the English word 'insect', which first appears in Philemon Holland's 1601 translation of the encyclopaedic "Historia Naturalis" written by the Roman naturalist Gaius Plinius Secundus (23-79AD) better known as Pliny the Elder (FIGURE 11.) nearly 2000 years ago. This massive work written by Pliny around 77AD is the largest Roman written work surviving to the present day and consists of 37 individual books on numerous aspects of the natural world including topics as diverse as astronomy, biology, geology and physiology. Despite living in an era long before the advent of modern scientific technology Pliny the Elder is widely regarded as a forerunner of scientific investigation. He was perhaps the first to provide a detailed account of the segmented anatomy of insects including wing structure as well as providing a description of the abdominal stingers and biting mouthparts that are a feature of several insect groups. Among many other subjects Pliny is also responsible for what is probably the first description of insect metamorphosis.

The topic of metamorphosis, with its connotations relating to rebirth and transition between worlds is a recurring spiritual theme that has influenced several cultures and their view of specific groups of organisms. The folklore surrounding many insects exemplifies some of the mythology that surrounds many of these animals today.

FIGURE 11. **Pliny the elder**, Gaius Plinius Secundus (23-79AD) better known as Pliny the Elder, Roman pioneer of scientific investigation. He is responsible for early detailed accounts of insect anatomy and metamorphosis.

Insect classification contains some anomalies, such as the inclusion of wingless insects such as fleas and lice in the otherwise winged Pterygota. The reason for their inclusion in the Pterygota despite their lack of wings relates to their presumed evolution from winged forms. In other words, they are regarded as being 'secondarily wingless'. Pterygota is currently included under the Supraorder Neoptera (TABLE 2), sometimes referred to as Martynov Neoptera following its initial use in 1923 by Russian entomologist and palaeontologist Andrey Martynov. The term Neoptera (derived from the ancient Greek 'New Wing') was first applied to insects characterised by their ability to fold their wings neatly over their abdomens. Another feature of insects belonging to the Pterygota is their division into two 'Superorders'; Endopterygota and Exopterygota (TABLE 2). These groups are distinguished from each other by differences in their developmental sequence. Members of the Endopterygota undergo a transition through a full metamorphic sequence including larval, nymph and pupal stages. However, Exopterygota have no pupal stage; their nymphs closely resemble small adults and simply transition to the adult form through a series of moults.

Pterygota currently include over 95 per cent of all insects, Orders not included in the Pterygota are the Archaeognatha (jumping bristle tails) and Zygentoma (silverfish and firebrats), which are now regarded as monophyletic organisms that evolved from primitive Arthropods dating from the Cretaceous era more that 320 million years ago.

# Order Coleoptera (beetles).

## Dung Beetles The world's strongest animals?

There are over 7000 species of dung beetle worldwide found chiefly in southern Europe, north Africa and western Asia. Most belong to the subfamily Scarabaeinae. These beetles lay their eggs in balls of dung collected from herbivorous mammals ranging from sheep and cattle to elephants depending on the species of dung beetle and their geographical location. The ball of dung provides a protective shield for the beetle eggs while it is rolled to a site selected by the parent beetle as a suitable location for the dung-protected eggs to be buried prior to the hatching of young beetles.

The magical emergence of new beetles from the buried dung was interpreted by the ancient Egyptians as a form of rebirth. As a result of this behaviour the Egyptian Scarab beetle *(Scarabaeus saser)*, was endowed with spiritual significance and was assumed to have special knowledge of the cosmos. This notion of spiritual enlightenment translated into a perceived relationship between the Scarab beetle and Khepri the ancient Egyptian god of morning sun. Khepri was seen as a manifestation of the sun god Ra and the rolling of the ball of dung by the beetle was interpreted by Khepri worshipers as representative of the passage of the sun across the sky.

Khepri is often portrayed in human form with a Scarab beetle head or as the beetle itself holding the morning sun (FIGURE 12.) The spiritual significance of the beetle was recognised by the ancient Egyptians in several ways, including the role of Scarab amulets[4] as part of a funeral service. During the funeral ceremony the scarab amulet is laid over the heart

---

[4] An amulet is a crafted object or trinket worn or carried as a charm as protection against evil influences.

of the deceased to ensure their safe passage into the afterlife, their protection from adverse judgment and their resurrection in the afterlife.

FIGURE 12. **Khepri, the ancient Egyptian god of the morning sun** is depicted with a Scarab beetle head or as the beetle itself holding the morning sun.

In addition to the mythology relating to Scarab beetles, these insects have also garnered 'real-world' fame such as their accreditation with record-breaking strength associated with their dung rolling and burial activity. On a 'weight for weight' basis dung beetles are some of the strongest animals in the world. Among their recorded feats is their remarkable capability of rolling or pulling many times their own weight (FIGURE 13). Perhaps the record goes to the aptly named bull-headed scarab (*Onthophagus taurus*), a native of several Mediterranean countries, which has been shown to be capable of pulling a ball of dung over a thousand times its own weight!

FIGURE 13. **Dung Beetles**, members of the genus *Onthophagus*, are capable of rolling and burying balls of herbivore dung many times their own weight.

## Ladybirds.

Another beetle with a strong spiritual heritage is the ladybird, which is probably better known as the ladybug in several countries including the United States.

> Although commonly used for insects the term 'bug' has been adopted for many arthropods and 'creepy-crawlies' such a centipedes, lice, mites and spiders. Anyone who has suffered from a 'stomach bug' will recognise that 'bug' extends to bacteria as well! True bugs are indeed insects, but only those belonging the order Hemiptera.
>
> Purists who insist on 'ladybird' instead of 'ladybug' might be countered with the observation that the 'bird' in 'ladybird' probably only arose because these beetles are able to fly like a bird. What's in a name?

The word 'lady' that forms part of the common name for this small red flying beetle refers to the Virgin Mary. The German name for the insect even translates as "Mary's Beetle". This might be a surprise to those who normally associate the colour blue with the costume worn by Mary in most western Christian artwork. However, the dominance of either red or blue in Mary's dress has been a topic of seemingly endless debate over several hundred years. Often both colours appear in pictures of Mary. Several eastern icons have the 'Mother of God' in a blue dress with a red mantle. However, depictions of Mary as 'Our Lady of Perpetual Help' in a red robe with a blue hooded mantle are often found in Roman Catholic and Eastern Orthodox iconography. Classical artists Botticelli, Da Vinci and Raphael picture Mary in a red tunic topped with a blue mantle. However, Mary dressed primarily in red appears in several frescoes and paintings of the renaissance era.

Irrespective of which colour, red or blue, dominates the artwork picturing 'The Virgin Mary, Mother of God', there is a general consensus within western Christianity that the colour blue is meant to depict the piety, purity and divinity associated with Mary's spiritual persona, whereas the red colour of Mary's clothing signifies her affinity with the earthly world including the blood and pain of Christ's passion. In this context Christian iconography has placed the bright red Ladybird insect under divine protection thereby giving the insect special powers. The bright red colour of the ladybird is therefore regarded as a lucky life-affirming colour common to many cultures. In the non-spiritual world, the powers attributed to the ladybird range from symbols of general good luck to the solution of earthly questions such as marriage and child-bearing prospects.

However, in the agricultural and garden world, any 'good luck' associated with the ladybird beetle and its positive effect on plant growth has a much more practical explanation. Aphids are small sap-sucking insects that can cause extensive damage to a variety of plants by feeding on leaves and stems, often in very large numbers. Ladybirds are natural aphid predators that can been introduced to aphid-infested plants as a means of controlling aphid numbers (FIGURE 14.). Ladybirds themselves cause little damage to plant material. Other insect predators that can be used for aphid control include damsel flies, hoverfly larvae and adult and larval lacewings.

FIGURE 14. **Ladybirds** are small (2-10mm) beetles of the family Coccinellidae. (order Coleoptera). Delicate flying wings are protected by hard colourful forewings that provide a dome-shaped protective shield after the beetle lands. They feed on aphids and some fungi.

# Order Orthoptera

## Cricket Anyone?

Other harbingers of good luck and good fortune in some cultures are crickets, which together with katydids, form the suborder Ensifera in the order Orthoptera. Grasshoppers and locusts comprise the other major orthopteran suborder Caelifera. The order Orthoptera was first described by the French entomologist Pierre André Latreille (1762-1833). However, in Chinese culture crickets have been regarded as omens of good luck happiness and prosperity for more than 2000 years. Nevertheless, the folklore attached to the cricket and its relatives is not always happiness and light.

The Orthoptera order comprises more than 20,000 species globally. The name may be translated as 'straight-winged'. Grasshoppers, crickets and their relatives are noted for their ability to neatly fold their wings, earning them membership of the infraclass Neoptera. Many orthopterans are characterised by their ability to produce a chirping sound by rubbing their wings or legs together. The appendages have corrugated patches that are responsible for the tone and frequency of their characteristic sound which acts as a signal to other organisms. It is this sound, some have called it a song, that underpins the folklore attached to these insects. In several European and American cultures the chirping sound that is characteristic of common cricket species such as the house cricket *Acheta domesticus*. (FIGURE 15) has been interpreted as a symbol of good luck, happiness and prosperity. In English folklore the cricket's song has signified the reincarnation of loved ones and the presence of protective household spirits. In contrast, a subsequent lengthy silence has been seen as an omen of death and misfortune. Worse, the killing of crickets is regarded as the precursor of terrible luck leading to disaster. Caution and respect for crickets by native American tribes such as the Hopi and Navajo have led to the incorporation of images of crickets into cultural rituals as a protective measure against the ravages of fate. In some respects, it is difficult to avoid an element of appeasement in some of these ceremonies, a sense that the cricket needs to be placated in order to avoid disastrous consequences.

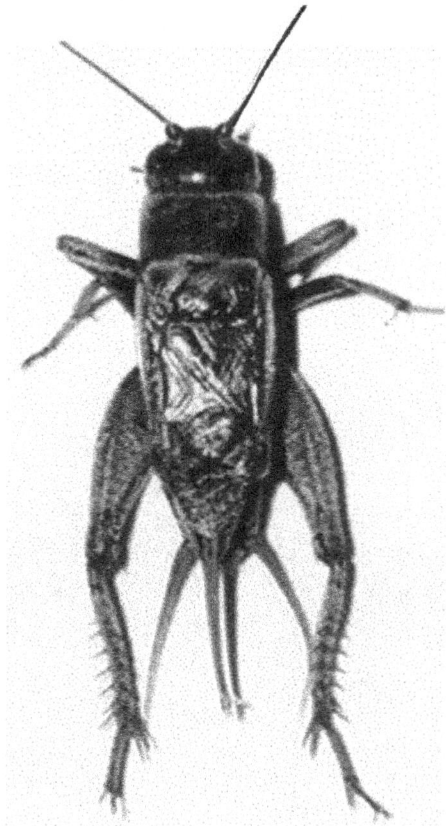

FIGURE 15. **House Cricket.** *Acheta domesticus*. Noted for its characteristic chirping sound thought to bring good luck.

The reputation of crickets as sources of good or bad fortune is shared by many different taxonomic groups including Katydids, which are also known as bush crickets. Katydids have also inherited the name 'long-horned grasshopper', which is a reference to their unusually long antennae, although their appearance is quite different from most grasshopper species. Indeed 'appearance' is perhaps not the right word to apply to these remarkably well-hidden orthopods. Katydids, which are the sole members of the Family Tettigoniidae, are primarily tree-dwellers that epitomise the reversal of the well-known phrase "seen but not heard". In contrast the descriptor "heard but not seen" is much more appropriate for these masters of disguise. The secret of their invisibility is their nocturnal habitat and their remarkable ability to disguise themselves as the leaves of trees that are their normal habitats. Apart from their leaf-green colour they are also capable of creating a remarkable impression of leaf veins (FIGURE 16).

FIGURE 16. **The Katydid**. *Tettigonia viridissima*, the only living members of the Orthopteran Family Tettigoniidae. Tree-dwellers primarily found in North America; capable of the extraordinary ability to disguise themselves as leaves, including very convincing copies of leaf veins.

## Grasshoppers.

Most grasshoppers belong to the Caelifera (FIGURE 17), a suborder of the Orthoptera with an evolutionary history going back nearly 300 million years to the early Triassic era, before the age of the dinosaurs.

The largest family within this suborder is the Acrididae whose members are characterised by their short stout antennae which are responsible for the common group name "short-horned grasshoppers".

FIGURE 17. **Red-legged Grasshopper.** *Melanoplus femurrubrum.* Order Orthoptera, suborder Caelifera, Family Acrididae. Commonly found in Canada, Mexico and U.S.A. A serious pest on cereal and leguminous crop plants. Adults grow to 1.5 – 2.5 cm in length

## Locusts.

Despite having a different common name, locusts are really a species of short-horned grasshopper belonging to the family Acrididae. Although these insects are solitary under most climatic conditions, they are capable of extraordinary swarming behaviour triggered by a surge in vegetative growth brought about by heavy rainfall following drought conditions. Although there are approximately 2,400 genera of grasshoppers comprising nearly 12,000 species there are only about 20 species of locust capable of this swarming behaviour. Although these are widely distributed throughout the world, probably the best known is the desert locust *Schistocerca gregaria* whose species name gregaria gives an immediate clue as to is swarming (gregarious) behaviour. A massive outbreak of swarming by this species occurred between 2003-2005 covering much of North Africa and several Middle East countries (FIGURE 18). Estimates of crop damage caused by this infestation exceeded $2.5 billion. Locust migrations associated with this swarming behaviour may exceed several thousand miles and may involve huge numbers of insects. In 2013 the island republic of Madagascar suffered an infestation of the migratory locust *Locusta migratoria* estimated to exceed one billion insects.

The discovery in Iran of a fossilised locust has traced the evolutionary history of locusts as far back as the early Oligocene era at least 30 million years ago. However, human records of locusts come from a variety of sources including tomb carvings from ancient Egypt dating from 2500-2200 BCE and ancient Egyptian murals around 1420 BCE. Further references to locusts appear in the Quran in the 9$^{th}$ century BC, in Homer's Iliad ca, 650 BC and in the biblical New Testament, where John the Baptist is sustained in the wilderness by eating locusts and honey

FIGURE 18. Locust swarms were a hazard in Ancient Egypt and remain an occasional problem today.

# Order Mantoidea.

## Mantises. Masters of Disguise.

Mantises are members of the insect order Mantodea comprising approximately 2,500 species divided among about 460 genera and 33 families. They are characterised by their typical 'praying' stance featuring fore-legs that are modified for feeding on insect prey.

Although mantises are often confused with stick insects (Phasmatodea) and other Orthopterans such as crickets and grasshoppers, members of this group of insects are now classified as part of the superorder Dictyoptera in the Order Mantodea. The order was so named by the German entomologist Karl Burmeister in 1829 with reference to the ancient Greek word mantis meaning prophet. The insect was given supernatural powers in both ancient Greece and ancient Egypt, and the religious undertones embodied in the species name of the European Green Mantis *Mantis religiosa* (FIGURE 18) also reflect the supernatural powers have been aligned with the characteristic praying pose that is responsible for the insect's common name 'praying mantis'.

Despite what is sometimes interpreted as its intimidating pose and a stick-like body that may grow up to 7cm in length the praying mantis poses no threat to humans. Its raptorial forelegs lined with spines enable the mantis to ambush small food insects that are captured by a 'lightning-fast' strike. The ambush is enhanced by the green colour of the mantis that disguises it as part of the plant visited by the prey insects.

FIGURE 18. **European Green Mantis** *Mantis religiosa*. An ambush predator of insects aided by its green colour and shape that allow it to hide on plants in wait for insect prey.

# Order Hemiptera.

Hemiptera is a large order of insects usually known as true bugs. They comprise more than 80,000 species including aphids, bed bugs, cicadas and leafhoppers.

The Hemiptera are now divided into the Sub orders Heteroptera, comprising what are now regarded as 'true bugs' (shield bugs, water bugs, assassin bugs, bed bugs), Auchenorrhynca (Cicadas) and Sternorrhynca (Aphids, Whitefly, Scale insects).

Members of the Hemiptera are generally characterised by having two sets of wings, often referred to as 'half wings'[5]; a membranous pair and a hard leathery pair. They also possess a sucking/biting set of mouthparts known as a rostrum which may be used to suck fluid from their respective food substrates. These might include leaf sap for the herbivore members of the order or blood for the carnivorous side of the order. For example bed bugs are capable of delivering a painful bite!

## Green Stink Bug.

The Green Stink Bug *Chinavia hilaris* (FIGURE 19.) is not generally harmful to humans or their pets and livestock. However, it is regarded as a nuisance pest of economic importance in the eastern United States and parts of the western U.S.A. where it feeds on the seeds and foliage of fruit trees, grape vines and crop plants such as tomatoes, beans, peas, cotton, and soybeans.

1-2 cm in length, its green colour can hide it from predators as it feeds on vegetables and fruit tree leaves. However, if

[5] Hence the name Hemiptera derived from the Latin for 'half wing'

discovered and disturbed it can discharge a foul-smelling exude to deter potential predators; Hence the name Stink Bug.

FIGURE 19. The **Green Stink Bug** *Chinavia hilaris*

# Order Hymenoptera.

## Bees, Wasps and Hornets.

Despite their armament in the form a painful sting in the tails of many bees and their relatives such as wasps and hornets, they have a relatively benign reputation with strong elements of wisdom, spirituality and industry. Much of the credit for these traits relates to the social behaviour of many of these insects, which can involve a good deal of teamwork. and which can include the sharing of quite complex information within their communities. This information can result in what might be termed 'division of labour' involving the delegation of different tasks within the bee community, usually depending on age and the immediate needs of the colony as a whole.

For example, young Honeybees a few days old have the task of cleaning the nest and preparing for nectar storage, whereas older bees up to about ten days old adopt what is termed a 'nursing' function which includes the feeding of young larvae and serving the needs of the queen bee. Between two to three weeks the community members turn into workers tasked with wax comb construction and food/nectar storage. After three weeks the bees become foragers and are also responsible for the defence of the nest. The reproductive function is monopolised by a single queen bee while infertile females become part of the worker bee community. This age-based 'employment model' known as 'Temporal Polyethism' has a degree of flexibility that can be activated by the needs of the bee colony as a whole. For example, a need for more foragers for the community is met by the survival of greater numbers of bees and their subsequent graduation into the mature foraging cohort.

Perhaps the most complex behavioural sequence performed by honeybees is the 'waggle dance'. This remarkable activity provides a means of communicating critical information about the location of an important resource from a single foraging honeybee to other members of its community. The resource of interest may be either the location of food or the whereabouts of a suitable nesting site. The dance is performed by a foraging individual in front of an audience consisting of several members of the forager's community and incorporates components that indicate both the direction and proximity of the desirable resource or 'prize'.

For example, the length and vigour of the dance is proportional to the <u>distance</u> of the prize from the performing honeybee and its audience. The <u>direction</u> of the prize resource (i.e. a nectar-rich flower or potential nesting site) is encoded by an extraordinary piece of geometry that relates the angle of the dance to the direction of the sun. The plane of the waggle dance depends on the habitat of the performing species. For cavity-nesting honeybees such as the western honeybees *Apis mellifera* or *A. nigrocincta*, the angle of the waggle dance indicating nectar-bearing plants is subtended in a vertical plane directly in line with the sun. (FIGURE 20.). However, some open-air nesting honeybees such as the Black Dwarf honeybee *Apis adreniformis* tend to pick a horizontal surface directly above their nest for their resource-indicating dance.

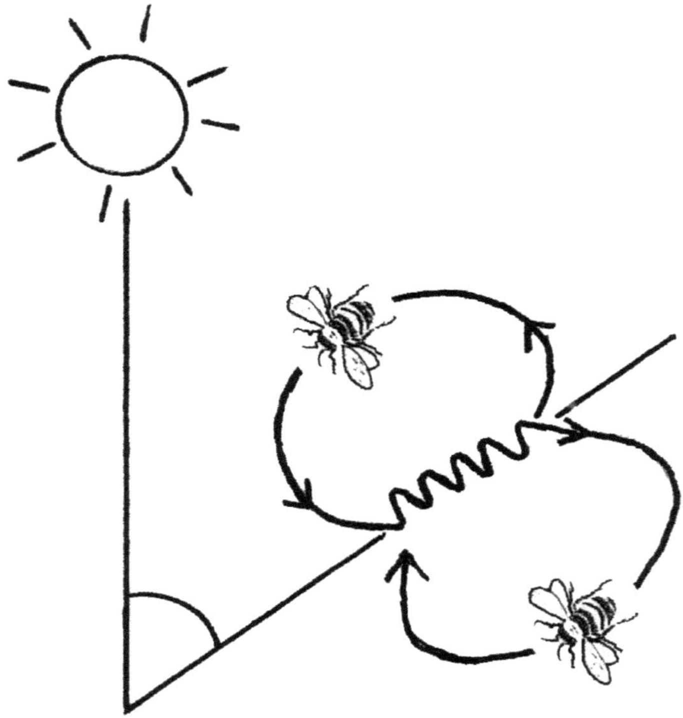

FIGURE 20. **The Honeybee Waggle Dance.** The critical angle relative to the sun's position is created by the diagonal (central) part of the figure 8- shaped dance. This is the section of the dance characterised by the back and forth waggling of the bee's abdomen. The frequency of this back and forth motion is an indication of distance to the resource.

Interestingly the waggle dance performance in front of other members of the bee's nest can be a 'two-way-street' for information exchange. For example, the presence of a dead bee on or near the target food source can have a restraining effect on the waggle-dancing forager as community members are able to signal danger among themselves and to the waggle dancer. As might be imagined, this extraordinary behaviour has been the subject of much historical interest including philosophers who have sought to delve into the cultural and spiritual implications of their complex social behaviour. In ancient Greek philosophy bees were closely linked to Socrates and his pupil Plato. In his book *Phaedo* Socrates refers to bees in his discourse concerning the immortality of the soul. It may (or may not) surprise philosophical scholars that the Greek student of Socrates, Aristotle (384-322 BCE), took an interest in this flower-seeking behaviour in honeybees and was seemingly aware that some sort of communicative behaviour accompanied this activity even if he was unaware of the exact nature of the dance. The tail-wagging dance of the honeybee was not deciphered until about 2,300 years after Aristotle's time. Decoding of the bee's waggle dance was performed by German Austrian ethologist Karl Ritter von Frisch in 1973. The discovery earned von Frisch the Nobel Prize for Physiology/medicine together with Nikolaas Tinbergen and Konrad Lorenz.

Not surprisingly a rich mythology surrounds the honeybee. According to the Roman statesman Cicero, the infant Plato was granted poetic insight as a result of bees settling on his lips. Ancient Egyptian, Greek and Roman ideology related to the honeybee contributed to the notion of bees as intermediaries between humans and the natural world which extended to the role of honey as an offering to the gods. Further west the sacred nature of bees became an integral part of Celtic culture, with Welsh poets declaring Britain as "The Isle of Honey". Welsh legend even stated that bees were born in Paradise and lived in the Garden of Eden where honey dripped from the trees. Given the rich cultural history and folklore associated with honeybees and their relatives it is of little surprise that these insects have been the inspiration for numerous works of art including poetry; for example, Rae

Howells' The Language of Bees (2022). Recommended reading also includes books by American Biologist and noted bee expert Thomas Seeley (Seeley 2010, 2025).

## A Different Sort of Honey Trap?

Speaking of honey, it is interesting to note that the complex tail-waggle dance of honeybees that is such a remarkable feature of their communicative skills may have a chemical as well as a physical component. Chemical analysis of female bees engaged in the waggle dance has shown that this activity is accompanied by the release of two compounds that have been described as 'communication pheromones'. These compounds are (Z)-9-Tricosene $C_{24}H_{48}$ and (Z)-9- Pentacene $C_{25}H_{50}$. Both are long-chain alkenes with a single double bond between the ninth and tenth carbon atoms.

(Z)-9-Tricosene (FIGURE 21) was first identified as a sex attractant produced by the female housefly *Musca domestica,* which secretes the compound to attract male flies. The compound also goes by the somewhat eponymous common name Muscalure. It might be questioned how a sex attractant might play a part in a physical activity such as the bee waggle dance whose goal is to communicate the position of a new resource to the bee community as a whole. However, when we remember that the resource being advertised by the dance may be a new nest, the very centre of honeybee sexual activity, then the role of the dance enhanced by muscalure becomes much clearer!

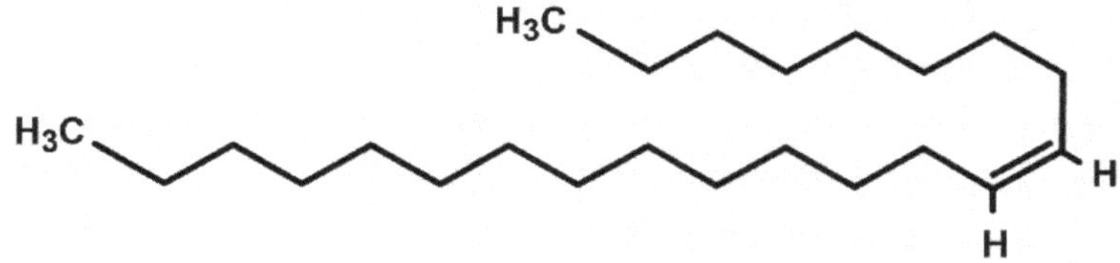

FIGURE 21. **Z-9-Tricosene**, otherwise known as Muscalure is a sexual attractant secreted by houseflies and honeybees to attract sexual partners.

The allure of Z-9-Tricosene has been noted by the pesticide industry who have devised a method of producing the compound from nervonic acid and oxygen using a process mediated by nicotinamide adenine dinucleotide phosphate (NADPH). Muscalure manufactured in this way has been used as a pheromone to lure male insects into traps to prevent them from reproducing.

In the human world the use of sexual allure to tempt dupes into traps or compromising situations has often been given the name 'Honey Trap'. Although Muscalure has no direct relationship with honey, the involvement of the honeybee (FIGURE 22.) in its production in nature adds an amusing irony to the notion of a 'Honey Trap' executed in this way!

FIGURE 22. **Honeybees** belong to the Genus *Apis*. The most common species is *Apis mellifera*.

As a postscript there are 270 bee species in the UK. Of these 24 are species of Bumblebee, which are also social, but don't make honey. Most of the remaining bees are solitary as are many types of wasps.

FIGURE 23 **BumbleBees** comprise more than 250 species belonging to the Genus *Bombus,* Family Apidae. They are not noted for honey production but are prolific pollinators.

## Ants.

All ants belong to the single family Formicidae in the order Hymenoptera. The Formicidae family is estimated to contain more than 20,000 species, of which about 14,000 have fully described and identified. A characteristic feature of ants that separates them from other members of the Hymenoptera is their jointed antennae, often referred to as geniculate or elbowed (FIGURE 24.)

FIGURE 24. **Common Black Ant.** Lasius niger. The most common species in Britain. Note the 'elbowed' antennae.

Ants are commonly found in all the continents in the world with the exception of Antarctica. They are also absent from Iceland and Greenland and reach the highest numbers in warmer climates such as tropical forests and savannas. Their colonisation of large areas relative to their diminutive size is a success story stretching back to their evolution from wasp-like ancestors from the Cretaceous era 150 million years ago. Unlike most other members of the Hymenoptera such as wasps and bees very few ants have wings. Winged ants are limited to reproductive forms, which at any time represent less than 5% of the total population.

Given their ubiquity throughout the globe, it is not surprising that varying estimates of biomass and total numbers of ants abound and are subject to regular updates as more is discovered about their worldwide distribution. A major obstacle to the accurate assessment of ant numbers is their small size and cryptic lifestyle. Although ants range in length between 2-20mm most are less than 6mm in length and only a tiny proportion reach the upper end of this size spectrum.

A further difficulty in tracking these insects is their hidden habitat which usually takes the form of subterranean nests. Numbers of ants in a single nest may vary from several hundred to several hundred thousand. However, some 'mega-nests' may even contain more than a million individuals.

Among the most recent estimates of the worldwide total number of ants is that of Schultheiss et al. (2022) who arrive at the extraordinary figure of 20 quadrillion, that is to say 20,000,000,000,000,000 individuals. This means that there are approximately 25 million ants for every human being on earth!

In terms of biomass the total biomass of ants worldwide exceeds that of all wild birds and mammals. When measured in terms of dry carbon the biomass of the global ant population, which is approximately 12 megatons of dried carbon, compares with a total global human biomass of 60 megatons. In other words, the total biomass of ants on earth is equivalent to 20% of the worldwide human biomass.

Most ant colonies are located underground although some 'canopy ant' species create arboreal nests. Colonies of ants are found in all of earth's major continents with the exception of Antarctica. A defining feature of ant colonies is their remarkable degree of organisation. The hierarchical ant society is dominated by (wingless) female worker ants that make up over 90% of ant populations. The fertile male population is restricted to a few drones, whose function is to fertilise the queen ant. Like other Hymenopteran 'queens' the queen ant is responsible for the propagation of the colony and as such may be its single indispensable member. In a single species of leaf-cutter ants *Acromyrmex heyeri* the male gender has even been dispensed with completely! This species consists solely of females that do not mate and reproduce by cloning themselves.

The essence of most ant colonies is 'teamwork', wherein the whole colony acts as a single entity, performing tasks that individuals would be incapable of, but which collectively benefit the whole colony. Examples include the collection and transport of food items, including prey, back to the nest by teams of ants. Such collective effort is magnified by the fact that some individual ant species can carry up to fifty times their own weight. This trait puts them among the strongest organisms in the animal kingdom.

Ants are omnivorous. Although most feed on vegetation several species such as red ants and fire ants have stings and mouthparts that can incapacitate or kill larger organisms. Some species of ant such as the 2cm long Bulldog Ant *Myrmecia pyriformis*, the Jack Jumper Ant *Myrmecia pilosula* have been known to attack humans with lethal effects. Members of the Myrmecia genus native to Australia are particularly recognised for their aggressive behaviour, which often involves the simultaneous use of a venomous bite and sting!

When predatory behaviour is coupled with mass teamwork the size of potential prey organisms can be vastly increased. This is often the case with red ants. Despite their small size, their ability to attack potential prey organisms in very large swarms (FIGURE 25) greatly increases their chances of incapacitating biota many times their size.

FIGURE 25. **The Common Red Ant** *Myrmica rubra* is native to Europe but has also been introduced to North America. They can deliver painful bites and stings, sometimes simultaneously, and will attack in swarms if their nest is disturbed. Note the larger queen ant.

## Termites

Termites (FIGURE 26) share many similarities with ants, such as their organisation into large colonies often reaching tens of thousands. Like ants, termite colonies can occasionally grow into super-colonies that number in the millions. Their body size is generally between 1-1.5 cm, which is similar to many of the larger ant species for which they are often mistaken. The difference in width of thorax and abdomen of termites is generally not as marked as most ant species; however the clearest distinction between ants and termites is the shape of the antenna, which is straight in termites but 'elbowed' in ants (compare FIGURES 24 and 26).

As is the case with ants, termite colonies are divided among several different castes including 'worker' and 'soldier' forms. Unlike ants, bees and wasps, where workers are exclusively female, termite workers comprise both immature males and immature females. When they mature into reproductive castes both male and female termites grow wings.

FIGURE 26. **The Drywood Termite** of the Kalotermitiditae Family tends to have a tubular abdomen and thorax that differs from the broad abdomen and relatively narrower thorax of ants. However, the straight antenna of termites compared with the 'elbowed' ant antenna is probably the clearest distinguishing characteristic of wingless stages of these insect groups.

Unlike flying ants, where the forewings are larger than the hind wings, the winged, reproductive termite castes, known as alates, have pairs of wings that are identical in size (FIGURE 27).

FIGURE 27. Unlike flying ants, the winged reproductive caste of termites has two pairs of wings that are identical in size. In ants the forewings in each pair are larger than the hind wings.

## Order Diptera (true flies).

Entomologists tend to refer to insects of this order as true flies based on the fact that the hind pair of wings have become transformed into appendages that are no longer used for flight and have become modified into club-like organs of balance known as 'halteres'. Dipterans therefore possess just one pair of membranous wings, i.e. a total of two wings responsible for the name 'Diptera' from the ancient Greek for di- two and pteron -wing. Diptera is a large order of over 150,000 species including horse flies, houseflies, hoverflies, mosquitoes and others. Dipterans are subdivided into three suborders: Nematocera. Brachycera and Cyclorrhapha.

Of the Nematocera, the best-known Families are the Culicidae (Mosquitos, Gnats), Chironomidae (Midges) and Tipulidae (Craneflies). Brachycera include Soldier Flies, Horse Flies and Gadflies. Cyclorrhapha include Hover Flies (Family Syrphidae), House Flies (FIGURE 28.) (Family Muscidae) and the well-known household pest, the Blue Bottle or Blow Fly *Calliphora vomitoria*.

FIGURE 28. **House fly** *Musca domestica* (Order Diptera, suborder Cyclorrhapha, family Muscidae) are characterised by large red eyes and a grey/black thorax with four longitudinal lines. They have a single pair of membranous wings plus halteres which assist in balance. They have no biting mouthparts and rely on the secretion of digestive enzymes to liquefy food item before ingestion.

## Screwworms.

Most blow flies feed on dead, decaying tissue from a range of plants and animals. They, therefore, pose little direct threat to healthy intact organisms including humans. However, they have the potential to accelerate the bacterial decay process thereby adding to the potential for bacterial infection. Exceptions are the parasitic blowflies *Cochliomyia hominivorax*, and *C. macellaria* which deliberately infest healthy i.e. undecayed animal tissue including that of humans. Although the targeted tissue may be fundamentally healthy these parasitic flies are drawn to animal tissues that have suffered cuts or abrasions. The flies deliberately lay their eggs in wounds so that emerging larvae (maggots) can feed on healthy undecayed animal tissue[6]. Female flies are each capable of laying up to 3000 eggs. It is the emerging parasitic larvae capable of burrowing into damaged skin that are responsible for the common name 'screwworm'. Apart from the tissue damage caused by the parasitic larvae themselves over their three-week life span, screwworm attacks also precipitate bacterial infections that are often lethal to the host mammals. Estimates of damage to livestock currently exceed $1.5 Billion. Screwworms are widespread in tropical and sub-tropical parts of the Caribbean and South America. Migration to the United States from Mexico was considered to have been eliminated in the 1960s, although the possibility of further outbreaks remains a topic of close vigilance for the U.S. Department of Agriculture.

Diptera undergo a complete metamorphosis including the formation of pupae and a series of larval stages. These are typically tubular with average lengths usually 1 cm, but which may reach 10 cm in some species. Wing pads are always

---

[6] The process of infestation of a live vertebrate by a maggot is known as *Myasis*.

absent in larvae but present in pupae.

Dipteran larvae are characterised by leg-less wormlike bodies that are found in a variety of habitats. These include burial in soil and mud, although some may be completely aquatic and are located at or close to surface water where aerial respiration in some species may be facilitated by air-breathing tubes extending to the air-water interface. Many larvae respire through the skin, and small gills are present in some species. Others obtain oxygen from the atmosphere using spiracles[7] and a few groups extract oxygen from plant tissues. The larvae of some Chironomid midges that live in anoxic habitats utilise a form of the respiratory pigment haemoglobin to enhance oxygen uptake. The pigment is responsible for the bright red colour of these larvae that are aptly called bloodworms.

## Transepithelial Potential.

Another interesting feature of aquatic dipteran larvae relates their ability to precisely regulate their internal ionic environment. This entails the maintenance of precise concentrations of major blood electrolytes such as sodium ($Na^+$), potassium ($K^+$), calcium ($Ca^{++}$), magnesium ($Mg^{++}$) and chloride ($Cl^-$) at levels well above those found in their natural (freshwater) environment. For example, the blood sodium level in a chironomid larva is more than 20X higher than the sodium concentration found in its normal freshwater environment. The question is, therefore. 'How is this difference between the internal and external electrolyte levels maintained?'

---

[7] Spiracles are small, paired openings on the bodies of insects, spiders and some other aquatic animals that regulate respiratory gas exchange. They can be opened and closed by valves.

In terrestrial vertebrates the principal organ for adjusting ionic and osmotic balance is the kidney. In adult insects the organ(s) that perform this function are tubular projections into the blood system from the mid- and hindgut known as Malpighian tubules.

Ionic regulation in the animal kingdom is an energy-consuming process known as active transport. Mechanisms that move electrolytes across membranes between body tissues and body fluids are called ionic pumps. These transepithelial or transmembrane pumps perform a variety of functions which may be excretory or may drive neuromuscular activity. Bearing in mind that electrolytes carry an electric charge ionic pumps may be responsible for producing transepithelial potential differences. i.e. the polarity on one side of a membrane or epithelium may be positive or negative relative to the opposite side of the membrane. For example, in neural cells at rest, the inside of the cell has a negative charge of -70mV relative to the outside of the cell. Large differences in $K^+$ concentration are also a feature of neural cells where intracellular $K^+$ is typically 120-150mM compared with extracellular $K^+$ levels of 3-5mM. Active pumps for different electrolytes may be linked and are sometimes depicted as 'revolving doors' wherein different cations such as $Na^+$ and $K^+$ are exchanged. Although such exchanges are often shown as one-to-one i.e. one $K^+$ in for one $Na^+$ out with each turn of the 'revolving door', things are often more complex than they initially seem. Detailed studies of cationic exchanges in neural cells often indicate a 'three $Na^+$ out' for 'two $K^+$ in' exchange resulting in a shift of the Transepithelial Potential (TEP). The influence of TEP in the transepithelial transport of ions has been given the name electrogenic ionic movement.

To investigate the interaction between transepithelial electrical potential (i.e. polarity) and the movement of electrolytes across membranes an experimental approach is required that measures ionic flux between 'internal' and 'external' epithelial compartments as well as determining the polarity and magnitude of transepithelial or transmembrane electrical potential difference. This brings us back to the dipteran larva!

In this respect the key characteristic of these aquatic larvae is that their osmo- and ionic regulation is largely confined to tiny papillae that protrude from the rear end of the larvae. These so-called anal papillae represent an important permeable "window" in what is otherwise an impermeable epithelium separating the larva's single body compartment from the external aquatic environment. These tiny ion-permeable windows (the papillae) are also the principal site of active transport mechanism(s) responsible for salt and water balance.

The highly localised position of the ionic and osmotic control mechanism in the anal papillae of the dipteran larvae enables access to both sides of the regulatory epithelium and facilitates the use of tiny microelectrodes capable of recording TEP as well as micropipettes for the ionic analysis of blood samples. An example of the experimental approach used to investigate the physiology of ionic and osmotic regulation is shown in FIGURE 29.

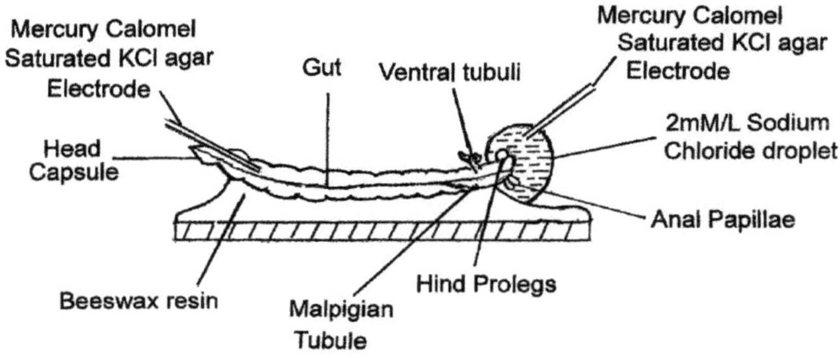


FIGURE 29. **Experimental preparation of Dipteran larva**, *Camptochironomus tentans* used to measure electrical potential across the larval epithelia. Larvae used for these experiment were usually approximately 2cm in length. Electrodes were replaced or supplemented by micropipettes for taking internal and external fluid samples for chemical analyses *Modified from Wright (1975).*

The 1960s were a period of major advances in understanding the ionic regulatory physiology of both terrestrial and aquatic animals including insects and crustaceans. Part of this progress was the increased availability of radioisotopes for experimental use. Isotopes such as sodium $(Na)^{22}$ enabled the tracking of electrolyte fluxes and concentrations in the tiny amounts of tissues and fluids that were available from small organisms such as insects and crustaceans.

Advances in analytical techniques such as atomic absorption spectroscopy (AAS) in the 1960s also contributed to this progress by enabling accurate analysis of body tissues and blood samples as little as 1µl in volume. This meant that detailed experiments on membrane physiology could be carried out on small organisms including arthropods that were available in large numbers and were often easy to culture in a laboratory.

Analytical tools such as AAS also extended to the measurement of industrial pollutants such as the trace metals cadmium, copper, mercury and zinc whose assessment by AAS provided a means for biomonitoring pollutants in both terrestrial and aquatic ecosystems. These studies often included the use of these organisms as prototypes for assessing the effects of environmental contaminants on biological systems such as neurophysiology and ionic regulation. Through studies such as these, scientists became aware of hitherto undiscovered relationships between trace metals and physiologically important ions.

For example, studies on calcified arthropods such as crustaceans have shown that cadmium ions $Cd^{++}$ are able to substitute for the calcium ions $Ca^{++}$ in transepithelial calcium pumps with the potential for interfering with the calcification of exoskeleton following each moult sequence. This is not to say that cadmium poses a threat to the calcification process under normal conditions, simply because of the relatively vastly higher calcium concentrations found in the natural environment. However, from an experimental perspective it is interesting to note that for some amphipod crustaceans the calcium pump responsible for post-moult recalcification has an even higher affinity for cadmium than for calcium. This is particularly evident in post-moult animals when the calcium pump is activated to recalcify the post-moult exoskeleton. If cadmium is present in the external medium its uptake and toxicity are, therefore, dramatically elevated in post-moult crustaceans.

At this point the reader might question what this has to do with 'the real world'? After all, cadmium is present in such low concentrations.

The questioner might also cast a sceptical eye on the revelation that the sodium pump that is an integral part of dipteran larval ionic and osmoregulation can be infiltrated by another metal ion, namely lithium. Experiments on these larvae have demonstrated that lithium can compete with sodium for access to the sodium pump. In its role as a monovalent cation $Li^+$ can also mimic the electrogenic characteristics of the sodium ion in influencing the transepithelial electrical potential.

> Lithium has been used as a drug since the 1870s when lithium bromide was first introduced as a treatment for gout and 'mania' (probably best described as a manic roller coaster). Published Research by Australian psychiatrist John Cade in 1949 and Danish psychiatrist Mogens Shou (1953) led to the use of lithium to treat bipolar disorder. This application was finally approved by the U.S. Food and Drug Administration in 1970.

As in the case of $Cd^{++}$ and $Ca^{++}$ it might be asked whether the $Li^+/Na^+$ relationship this be another case of a physiologically functionless metal ion ($Li^+$) competing with a critically functional metal ion ($Na^+$)? In the field of experimental physiology there are at least two different responses to this. The first is that, unlike cadmium, lithium <u>does</u> have a physiological function that is relevant to the human condition (see below).

Despite its successful use for treating bipolar disorder, little is known about the mode of action of lithium in human neurophysiology, although empirical clinical data warn against the potentially toxic effects of lithium 'overdose'. It is interesting that a significant factor in controlling the potentially toxic effects of lithium relates to the accompanying sodium dosage. Lithium consumers are advised to avoid low sodium or sodium-free intake when using lithium as a drug. This implies that sodium probably has a mitigating effect on the bioactivity of lithium; suggesting a competitive effect perhaps.

Of course, these revelations are a long way from the competitive effect of lithium on the sodium pump of dipteran larvae as described earlier. Some readers may find it surprising that the scientists who generated those data had no ulterior motive beyond the simple demonstration that such a competitive $Li^+/Na^+$ relationship exists, at least in dipteran larvae and 'probably other organisms as well'. Some may argue that such studies could have eventually led directly to a cure for bipolar disorder. Highly unlikely, but you never know! Science is not always concerned with heading for some shining light on the horizon. Nor does it always travel in straight lines.

Speaking from personal knowledge of a few of the non-linear thinkers responsible for some physiological tenets that still hold today, I feel that, sometimes originality and eccentricity go hand-in-hand. I recall one of my senior academic colleagues in the 1970s, sitting at his desk amid piles of books and papers. On his desk sat a 2' X 2' X 2' cubic wooden frame covered with fine cotton cheesecloth netting over five of the six 'faces' of the cube. The sixth 'face' consisted largely of a sleeve, also made of cheesecloth.

Our conversation consisted mainly of a discussion of results of recent experiments and was barely interrupted as my colleague calmly rolled up one of his shirtsleeves and inserting his arm into the cheesecloth sleeve. This activity was accompanied with the word "lunch", which seemed a little odd until it became obvious that 'he' (my colleague) <u>was</u> lunch. Inside the box-frame was his mosquito culture. Within a few seconds his bare arm was covered with mosquitoes! Perhaps the oddest aspect of this sequence was the normalcy of the conversation, which consisted of a couple of minutes of 'small talk' while his insects finished their meal on his arm. The removal of his arm from the sleeve was completed always without any escapees. I witnessed this performance many times without mishap. It should be added that the insect culture was disease free and my colleague was immune to any discomfort having spent his childhood in the native country of this insect. As for me, I was grateful that the cheesecloth didn't have any holes.

## 'Lord of the Flies'.

House Flies and many of their close relatives have been commonly branded as 'nuisances' or worse in human culture going back to ancient Greece. Pliny the Elder (FIGURE 11.) refers to Myiagros as a cult figure imbued with the ability to relieve pestilence by warding off flies, an ability that carried heroic status worthy of animal sacrifice. There are some accounts of sacrifices being made to the flies themselves, or to a version of the god Zeus who was credited with the ability to shoo away flies. However, it seems that the power of Zeus exceeded simply the ability to shoo away flies. According to Greek mythology surrounding Bellerophon, the Greek hero, in a fit of hubris incurred the wrath of Zeus by attempting to approach Mount Olympus, the home of the gods, on his horse Pegasus. Zeus, clearly regarded this as a 'step too far' and sends a gadfly to bite Pegasus, causing the horse to rear up, pitching Bellerophon back down to earth. Further accounts of the impertinence of flies appear in several of Greek storyteller Aesop's fables. These tell of the mischievous behaviour of flies in their interaction with other animals including humans. Some fables have 'moral' themes, such as the story of the Gnat and the Lion. In this tale the gnat, bragging of its superiority after biting the lion meets its downfall by being caught in a spider's web. The moral? 'Pride comes before a fall'.

Beelzebub, embodying gluttony and known as one of the seven Princes of Hell or Seven Demons, was also known as the Lord of the Flies and was depicted as such in the 1863 edition of Jacques Collin de Plancy's "Dictionnaire Infernal" (FIGURE 30.). 'Lord of the Flies' was also the title of William Golding's 1954 novel concerning the emergence of savagery among boys shipwrecked on a desert island. Although the allegoric theme of novel deals with the breakdown of civilisation under difficult circumstances rather than outright demonology there is no doubt about the linkage between flies and wickedness.

The negative theme surrounding these insects continues in the 1957 science fiction novelette 'The Fly' by French-British author George Langelaan. Widely regarded as a horror story 'The Fly' tells the tale of an eccentric scientist whose experiment with a fly goes wrong and results in him turning into a grotesque Fly/Human hybrid. 'The Fly' formed the basis for the 1986 film of the same name, memorable for the famous line "Be afraid. Be very afraid".

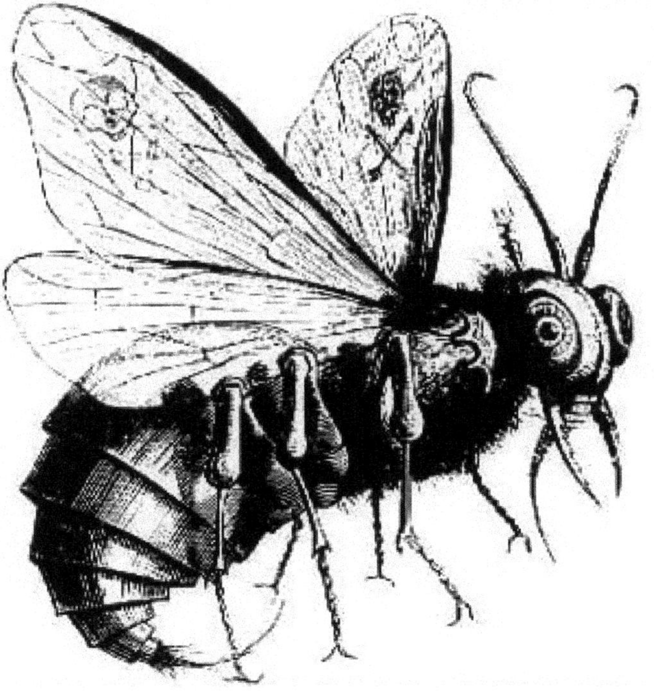

FIGURE 30. **Belzebut (Beelzebub)**, 'Lord of the Flies' as pictured by Louis Le Breton in the 1863 edition of Dictionnaire Infernale by in Jacques Collin de Plancy.

Despite the mischief and demonology attached to the fly, the insect has also been the subject of more benign folklore, including the poem '*The Fly*', first published in "*Songs of Experience*" in 1794 by English poet William Blake.

## **The Fly.** *by William Blake.*

*Little fly,*
*Thy summer's play*
*My thoughtless hand*
*Has brushed away.*

*Am not I*
*A fly like thee?*
*Or art not thou*
*A man like me?*

*For I dance*
*And drink and sing*
*Till some blind hand*
*Shall brush my wing*

*If thought is life*
*And strength and breath,*
*And the want*
*Of thought is death,*

*Then am I*
*A happy fly,*
*If I live,*
*Or if I die.*

# Subphylum Myriapoda

## Order Polyzonida

### Millipedes.

Most millipedes are worm-like organisms in the Subphylum Myriapoda; Order Polyzonida, with long cylindrical or flattened bodies and multiple body segments (more than 20 overall). Most millipedes belong to the Family Siphonotidae which comprises more than seventy species divided among 13 different genera.

The name millipede derives from the Latin words 'mille' (thousand) and pes (feet), and stems from the anatomical feature wherein each body segment generally has two pairs of jointed legs, a characteristic that is also responsible for the class name Diplopoda (two feet).

The Diplopoda is the largest class of the Myriapoda which also includes the centipedes and other multi-legged organisms. For many years the names millipede and centipede were used as shorthand for 'a lot of legs'. However, as recently as 2021 Marek et al. finally described a millipede from West Australia that really did live up to the name millipede. Hence the title of the paper "The first true millipede – 1,306 legs long". The revelation also extended to no less than 330 body segments!. The name of this new ultralong species is *Eumillipes Persephone* (FIGURE 31)

FIGURE 31. *Eumillipes Persephone* (Marek et al. (2021). Millipede with the largest number of legs

Although *Eumillipes Persephone* is among the largest of living millipedes and may hold the record for the most legs, perhaps the largest existing species is the giant African millipede *Archspirostreptus gigas*, which can grow up to 38cm (15") in length. However, even this species would have been dwarfed by some prehistoric specimens. Fossil evidence suggests that the extinct millipede group Arthropleura from the late Ordovician era more than 400 million years grew to over 2m (6.5').

Millipedes are harmless to humans and other animals although some are regarded as garden pests as a result of their consumption of sap from some vegetable seedlings.

# Subphylum Crustacea.

The Crustacea is a subphylum of the phylum Orthoptera comprising approximately 65,000 identified species. In 2010 crustaceans were included in the clade Pancrustacea which includes all crustaceans and all hexapods.

Crustaceans are primarily aquatic, inhabiting both marine and freshwater habitats, although a few are terrestrial and/or parasitic.

Crustaceans possess a segmented body that is divided into three primary regions: the cephalon (head), perion (thorax) and pleon (abdomen). In many species the head and thorax are fused to form the cephalothorax which often has a hard calcified protective covering known as the carapace. Each segment is protected by a hard exoskeleton which is periodically moulted to allow growth

Each body segment has paired appendages that perform a variety of functions including sensory, feeding and locomotion. In the Class Malacostraca, there are also abdominal appendages which aid swimming, respiration and reproduction. Abdomens in this group end in a telson which is often flanked by uropods to form a tail fan.

# Class Remipedia.

This class of multi-legged millipede-like organisms, first discovered in 1981, are considered to be members of the Pancrustacea Clade and as such they are identified as a Class within the subphylum Crustacea. This class comprises 8 Families 12 genera and 28 species. They are aquatic in habitat and tend to be found primarily in salt water. Their paddle-shaped appendages are responsible for the Class name Remipedia which translates as oar-footed.

Perhaps the most notorious of this class, and possibly the Crustacean subphylum itself is the species *Xibalbanus tulumensis* (FIGURE 32) which has the distinction of being the only venomous member of the Crustacea. This blind cave-dwelling species was discovered by caver and biologist Jill Yager in 1987 in an underwater cave on the Yucatán Peninsula in the Caribbean Sea. (Kaplan 2013; von Reumont et al. (2013).

The venom produced by *X. tulumensis* has been identified as a mixture of chitinases and peptidases used to hydrolyse proteins and liquefy small prey organisms prior to their ingestion.

FIGURE 32. **Sole venomous Remipede Crustacean**
*Xibalbanus tulumensis.* discovered in 1987 in underwater caves on the Yucatán Peninsula in the Caribbean Sea

# Arthropods as Biomonitors.

The phylum Arthropoda accounts for a broad spectrum of animals that form an integral part of terrestrial ecosystems and numerous aquatic habitats ranging from freshwater and estuarine to deep ocean.

In these environments arthropods such as crustaceans and insects may be part of the food chain or, more importantly, may exert critical influence on food webs, as typified by the damage caused to food crops and livestock by insect pests. However, insects may also be beneficial to some aspects of agriculture where crop pests themselves are preyed upon by insects.

Some mysid shrimps are cultured in laboratories for experimental purposes and are often used as a food source for other cultured marine organisms. Because of their sensitivity to water pollution, these crustaceans are sometimes used as bioindicators to monitor water quality.

In several cases, the health and growth of arthropod populations may reflect the health of the ecosystems as a whole. Human interaction with these organisms has Inevitably focussed on their role as part of the food chain (e.g. edible crustaceans and insects) and, more importantly, their influence on food organisms important to humans and their livestock.

Where human pollution has a negative effect on natural or agricultural systems, field populations of Arthropods can provide us with important information on the degree of contamination and potentially negative effects. Observations of community dynamics can provide information on the effects of pollutants on reproductive performance whereas observations at the organismal level can be used as models that help us understand the mechanisms behind the toxicity of individual chemicals or chemical mixtures.

Among the simpler laboratory-based experiments used to test the toxicity of single chemicals or complex mixtures are so-called dose response assays. These consist of controlled exposures of test organisms to a range of concentrations of a specific chemical or a mixture of chemicals[8] that might come from an industrial effluent.

Arthropods are frequently used as test organisms in dose-response assays due to their ready availability from terrestrial and aquatic habitats and the relative ease of culturing these organisms in a laboratory. These toxicity tests are often carried out on insect and crustacean adults or larvae. Test animals are exposed to a range of chemical doses over a specific period that might be hours or weeks in length. Toxicity endpoints usually range from mortality to sublethal effects such as growth or reproductive performance e.g. number of viable offspring produced. FIGURE 33 shows a typical 'dose-response curve obtained by exposing organisms to a range of chemicals that determined by preliminary, range-finding tests to evoke responses ranging from zero (no measurable response), to a maximum (100%) response.

---

[8] Chemical mixtures may be prepared in the laboratory or may consist of a chemical effluent that might have unknown components.

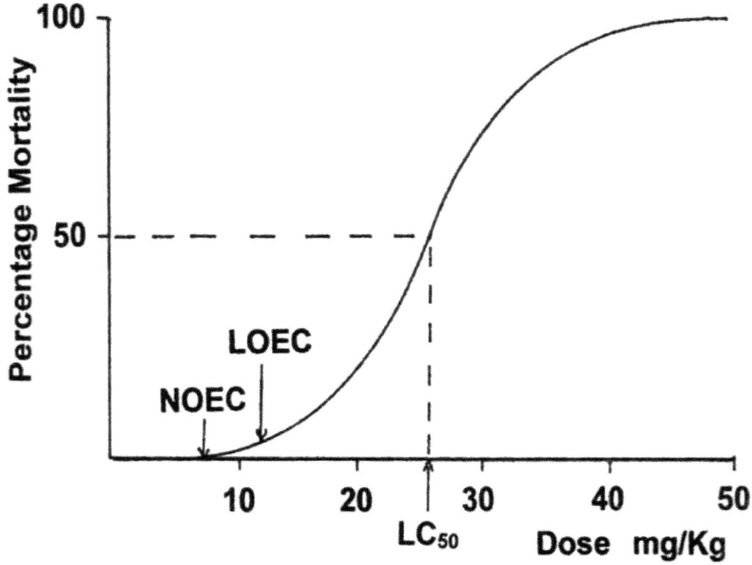


FIGURE 33. **Dose-Response** curve showing the relationship between the acute toxic endpoint Percentage Mortality and chemical dose. In sublethal assays 'Percent Mortality' may be substituted by 'Percent Inhibition' of e.g. growth or reproduction. NOEC (No Observed Effect Concentration) and LOEC (Lowest Observed Effect Concentration) are used to assess sublethal toxicity. LC50 is a standard measure of acute toxicity.

In a simple toxicity bioassay '100% response' is defined as the total mortality of the test population within the prescribed time course of the assay which could range from hours to days. In the case of a non-logarithmic (linear) range of chemicals the response (percentage mortality) follows an S-shaped curve, (FIGURE 33) indicating what is described as a 'normal distribution' of response. This shows mortalities of particularly sensitive individuals at the lower end of the curve, and the mortalities of the most resistant organisms at the upper end of the curve. Mortalities of the majority of the population are shown in the central portion of the curve. Other important parameters in the lower half of the curve are the 'Lowest Observed Effect Concentration' (LOEC) and the 'No Observed Effect Concentration (NOEC). As its name implies The LOEC is the lowest chemical concentration that causes a statistically significant response when compared with 'Control' animals i.e. those with no chemical exposure. By comparison the NOEC may have a chemical concentration that is statistically greater than zero but does not evoke a measurable response from the test organisms.

From the regulatory perspective LOEC and NOEC are important parameters in determining safe chemical discharge levels in an ideal world. For example, the geometric mean of the LOEC and NOEC has been adopted by regulatory agencies as a gauge of sublethal or chronic toxicity known as the Maximum Acceptable Toxic Concentration or MATC.

However, from a practical standpoint it is understood that chemical toxicity is an unchangeable and unavoidable hazard. It is, therefore, important to define a benchmark that expresses toxicity and that can be used for all chemical toxicants. This universal standard is the $LC_{50}$, which is defined as the Lethal Concentration that results in the deaths of 50% of the test population over the prescribed length of the toxicity bioassay. The $LC_{50}$ shown in FIGURE 33 is, therefore, a dose of 26mg/Kg.

Instead of a body dose such as mg/Kg the acute toxicity term $LC_{50}$ may also be expressed as the chemical concentration of the external medium. For example, in an aquatic environment the $LC_{50}$ be expressed as mg/L, i.e. the concentration resulting in 50% mortality of the test population. Although $LC_{50}$ is universally regarded as the standard means of expressing acute toxicity, there may be circumstances where it is important to estimate a chemical level resulting in a lower degree of mortality. Such toxicity estimates may, for example, be expressed as $LC_{25}$ or $LC_{10}$.

Another form of toxicity assessment may relate to an endpoint other than mortality. In such cases, the letter 'E'(Effect) in LOEC and NOEC would refer to a sublethal endpoint such as growth or reproductive performance. In these assays the endpoints (growth and reproduction) are directly compared with control (untreated) organisms measured concomitantly. In such assays the vertical Y-Axis 'Percentage Mortality' in FIGURE 33 would be replaced by a sub-lethal endpoint such as 'Relative Inhibition of Growth' compared with untreated controls. Where test parameters in treated test organisms showed no difference from controls, the percentage inhibition would be scored as zero. Conversely, lack of growth or reproduction in treated animals may be scored as 100% (inhibition) on the dose-response vertical axis. Using this approach it is, therefore, possible to measure a 'median effective concentration' or $EC_{50}$. This is analogous to the $LC_{50}$ and can be defined as the chemical concentration causing a sublethal response in 50% of the population.

Other variables that may affect acute or chronic chemical toxicity include ambient environmental conditions acting individually or in tandem. An example of this is illustrated by a series of experiments conducted on the toxic effects of organotin compounds on marine and other aquatic wildlife.

# The Rise and Fall of Organotins as Antifoulants.

In the 1960s 'organotin' compounds were developed as antifouling agents for worldwide application to the hulls of ships. Primary targets included encrusting crustaceans such as barnacles which were responsible for multimillion dollar increases in worldwide maritime fuel consumption resulting from the increased drag caused by their infestation of ships' hulls. These antifouling compounds came to be known by the generic term 'Tributyltins' (TBTs), although this term was applied to group of closely related chemicals. Of the TBT group of compounds, the most commonly used was tributyltin oxide (FIGURE 34).

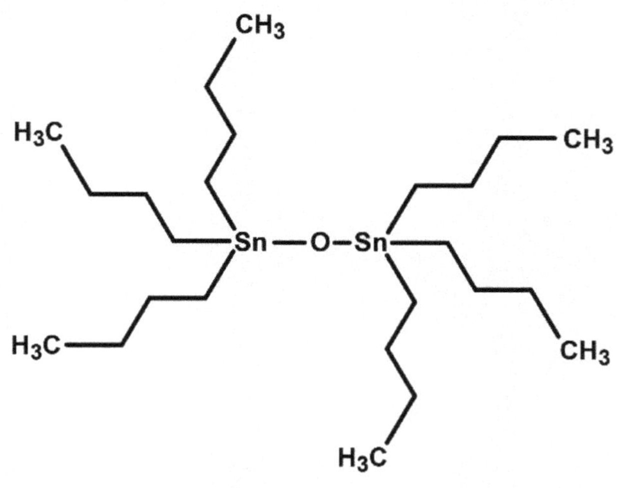


FIGURE 34. **Tributyltin oxide** was the primary component of antifouling paint first used in the late 1960s but were finally banned 40 years later due to their toxicity to marine life.

Because of their antifouling properties, TBTs and their close relatives trimethyltins became the subject of several toxicological studies with particular focus on encrusting crustaceans such as barnacles. Other crustaceans also served as surrogates in assessing various aspects of TBT toxicity. One such study was directed towards understanding the various environmental conditions likely to affect organotin toxicity. In this study (Wright and Roosenburg 1982) the four variables temperature, salinity, TBT exposure time and exposure concentration were plotted in a three-dimensional format as shown in FIGURE 35.

In view of the complexity of the matrix involved the data presented in FIGURE 35 include only the highest and lowest of the salinities investigated, i.e. 10°/o and 30°/o. Although not reported here the response surfaces from 20°/o salinity registered intermediate responses compared with the 10°/o and 30°/o response surfaces. The overall conclusion from these studies was that TBT toxicity is exacerbated at lower salinities and higher temperatures.

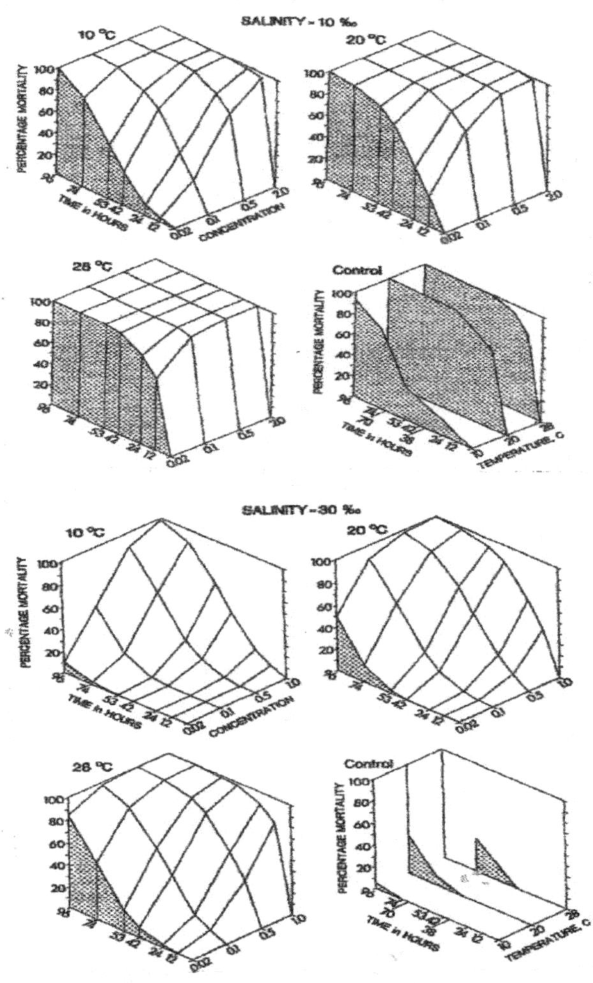


FIGURE 35. **Response surfaces** showing effect of temperature on trimethyltin toxicity to larval crustacean *Uca pugilator* at salinities of 10‰ and 30‰

When dealing with complex multivariate matrices such as these, large numbers of test organisms are needed to generate statistically reliable results. It must also be borne in mind that the data points used to construct the response surfaces presented here are the means from multiple individuals. It is, therefore, easy to see the need for large numbers of conspecific, and preferably cultured individuals.

The importance of larval insects and crustaceans as a reliable and easily accessible source of test biota has long been recognised. Particular significance might have been attached to the use of crustaceans as organisms for testing TBT toxicity in view of their taxonomic affiliation with (crustacean) barnacles which are the main source of shipboard fouling. However, the toxicology that finally decided the fate of TBT as an antifouling agent came from a quite different, and unexpected direction.

By the mid-1960s TBT had become the world's most popular antifouling agent and had been applied to a significant proportion of world shipping. In addition to its effectiveness as an antifouling agent TBT also had antifungal properties, which made it useful as a wood preservative.

However, by the 1980s is had become apparent that TBT compounds that sloughed off from ships' hulls had endocrine disrupting properties that affected the sexual development of a wide range of aquatic organisms within the vicinity of treated vessels. In many animals the term 'imposex' was used to describe a TBT-induced condition causing the abnormal development of sexual organs and consequent impotence (Bettin et al. 1996). The lipid soluble organic component of the TBT molecule and related compounds contributed to their bioaccumulation in lipid tissue and their biomagnification throughout the food chain. This meant that the imposex condition began to appear in mammalian consumers at the top of the food chain such as seals, whales and dolphins. This condition was coincident with high levels of TBT in exposed organisms. TBT also caused a range of immunosuppressive effects with a variety of symptoms. In humans prolonged TBT exposure was linked to a numerous health problems, including headaches, fatigue, respiratory issues, and more. Long-term TBT exposure was also found to cause liver and kidney damage.

As a result of the multiple adverse effects of TBT its ban by the U.N. Marine Environmental Protection Committee was initially advocated in the 1980s. This led to a worldwide ban on organotin applications on ships starting in 2003. By 2008 organotin compounds acting as biocides were completely banned from use in anti-fouling paint under the Rotterdam Environmental Convention and have now been banned by the International Convention on the Control of Harmful Anti-fouling Systems on Ships of the International Maritime Organization. As recently as 2018 criminal indictments were attached to the continued illicit use of TBT-based paints as antifouling agents.

Having viewed TBT and similar chemicals as subjects of numerous international regulations, and marine environmental conventions it is interesting to turn, once again, to the insect order Diptera as a source of research data providing important information on the mode of action of organotin compounds. Papers by Vogt et al (2007) and Lilley et al. (2009) are among several publications documenting the use of larvae of the midge *Chironomus dorsalis* as models for assessing organotin toxicity at the genetic and population levels.

In several respects both adult and larval members of the Arthropoda have proven to be useful models for improving our understanding of chemical and other environmental stressors on ecosystems.

# Worms as Biomonitors and Decomposers.

Another group of biota with the potential for biomonitoring are members of the Annelida, a phylum containing about 25,000 species including earthworms, leeches and ragworms. While leeches are primarily parasitic on other organisms, ragworms, also known as polychaetes are generally found in the marine environment, often in burrows close to the water's edge where they extract their food from the sediment. Evidence of their burrowing behaviour takes the form of piles of digested and excreted sediment in the seaside intertidal zone.

At first sight it is surprising that such creepy crawly organisms have found their way into the world of art and poetry over the last few hundred years. On the other hand, there is the possibility that their slimy nature and many legs may be a key part of their attraction!

In his epic (144 verse!) poem "The Rime of The Ancient Mariner" extolling the virtues and dangers of life aboard ships, Samuel Taylor Coleridge in 1798 wrote:

*Water, water, everywhere,*

*And all the boards did shrink;*

*Water, water, everywhere,*

*Nor any drop to drink.*

*The very deeps did rot: O Christ!*

*That ever this should be!*

*Yea slimy things did crawl with legs*

*Upon the slimy sea.*

FIGURE 36 shows an illustration by German biologist Mattias J. Schleiden (1804-1881) published in *Das Meer (The Sea)* in 1869. The plate depicts several members of the class Polychaeta which was first formally recognised by German biologist Adolphe Grube in 1850.

FIGURE 36. **'A Variety of Marine Worms'**. An illustration by German Biologist Mattias Schleiden published in Das Meer (The Sea) in 1869.

Terrestrial earthworms (FIGURE 37) show similar signs of burrowing activity in the familiar form of 'worm castes' that are commonly found in meadowland and gardens. This continual processing of soil and sediment means that these animals can provide a means of assessing the bioavailability of chemicals from the surrounding sand and mud. (Tischer 2009).

Earthworms have also been used to assess biological responses to specific environmental contaminants including cellular biomarkers of pollution-induced stress (Rutgers et al. 2009). Earthworm biomarkers used to assess soil pollution have included the degree of haemoglobin oxidation, changes in granulocyte (white blood cell) morphometry and count. Other biomarkers of soil pollution produced by earthworms have included the production of metal-binding proteins called metallothioneins[9] and changes in enzymes such as acetylcholinesterase. as indication of exposure and effect of xenobiotics in polluted soil environments.

A paper by Callahan and co-workers (1991) describes a suite of *in situ* bioassays conducted using earthworms at a Superfund site in Holbrook Massachusetts U.S.A. Superfund sites are designated by the U.S. Environmental Protection Agency (EPA) as exceptionally highly contaminated sites requiring a long-term clean-up plan. The Holbrook site had previously been used for preparing and mixing several of the pesticides that now contaminated soil at the site, and was designated by the U.S.EPA as hazardous, hence its Superfund status. Superfund sites typically include landfills, manufacturing plants and mines.

---

[9] Exposure to trace metals such as cadmium, copper and zinc may induce in biota the production of metal-binding proteins called metallothioneins. These proteins are protective in nature sequestering metals in a form that prevents them from binding to metabolically important proteins such as enzymes. The presence of metallothioneins may act as a bioindicator of the presence of potentially toxic trace metals.

An important component of the hazard assessment programme at the Holbrook location was the placement of batches of earthworms in a series of transects that were deliberately chosen to cover areas of high and low chemical contamination. Following a period of exposure to enable the worms to sample their environment stress endpoints recorded included mortality and clear signs of morbidity such as coiling, stiffening, swelling and the development of lesions. In addition to physical stress symptoms worms were also subjected to chemical analyses designed to assess the bioavailability of the chemical contaminants.

Apart from a variety of 'effects biomarkers' that may emerge from the analysis of earthworm tissue a more straightforward endpoint signifying soil quality is a simple earthworm count. Earthworm densities have long served as empirical indicators of soil quality. 'The more worms, the healthier the soil'. Apart from chemical contamination, soil texture is an important factor in this assessment of soil health, i.e. fine-grained soils favour many vegetables and small garden plants.

FIGURE 37. **The Common Earthworm** *Lumbricus terrestris*: useful for assessing the health of terrestrial substrates and providing information on the bioavailability of environmental contaminants.

Like other creepy crawly animals, even earthworms, have found their way into verse. The following poem (FIGURE 38) narrates the story of a young girl rescuing an earthworm from the clutches of a bird predator. The young worm saviour extols the virtues of the worm as an important asset to the soil community.

FIGURE 38. **The Earthworm poem** about a young girl who saves an earthworm from a bird predator, enabling it to continue enriching the soil.

**Earthworm Poem.** (Anonymous).

*This poem is slimy*
*And twisty, and squirmy*
*It tells the tall tale*
*Of a squishy earthwormy*
*Who lived underground*
*By munching old leaves*
*Turning them into soil for trees*
*The worm couldn't see*
*No eyes, arms, or feet*
*But helped plants to grow*
*Chew, poo, then repeat*

*One day a bluebirdy*
*Was out for a snacky*
*And saw the earthworm*
*What then happened was wacky*
*Bird flew for the worm*
*To eat it for lunch*

*But little girl, Lucy*
*Stepped in with a crunch*
*"Leave wormy alone*
*He might be muck-yucky*
*But the soil he makes*
*Makes all of us lucky!*
*I'll protect mister earthworm!*
*With all that I'm worth*
*We need slimy earthworms*
*They're good for the earth!"*

# Roundworms.

In view of the secretive and mysterious nature of many of the creatures featured in this book it seems appropriate to end with a phylum that is perhaps the most cryptic of all. Roundworms, also known as Eelworms, belong to the Phylum Nematoda, a taxonomic group that inhabits all parts of the earth's mantle and crust, and whose ancestry goes back to the early Devonian era 400 million years ago. The name Nematoda is derived from the Greek word translated as 'Nema' meaning thread or yarn, a term that aptly describes their thread-like appearance.

Nematode worms are primarily free-living creatures that are found in soils and sediments nearly everywhere on earth as well as most freshwater and seawater habitats including ocean deeps. It is not unreasonable to speculate that a random sample of substrate taken from almost anywhere on the planet has a good chance of containing at least one specimen of this taxonomic group. For example, approximately 90% of the animals on the ocean floor belong to the Nematoda.

It is estimated that the phylum Nematoda contains more than 2,200 genera divided among 256 families. However, some aspects of their taxonomy remain somewhat speculative due to their small size, often just a few millimeters in free-living forms, and a paucity of distinguishing physical characteristics among many the free-living forms. This degree of uncertainty has been responsible for massive discrepancies in species estimates. For example, although the formally recognised number of nematode species is in the region of 28,000, genetic sequencing, often referred to as DNA barcoding[10], has led to estimates of nematode species in the hundreds of thousands and even approaching a million.

Approximately one third of all nematodes are parasitic with a preference for mammalian hosts including humans. Despite retaining the thread-like form, parasitic nematodes tend to be much larger than free-living forms. For example, the most common human nematode parasite *Ascaris lumbricoides* may grow to 4mm in diameter and 30cm (12") - 50cm (19") long in male and female (FIGURE 39) worms respectively. The number of humans infected by this species has been estimated to be approximately one billion worldwide.

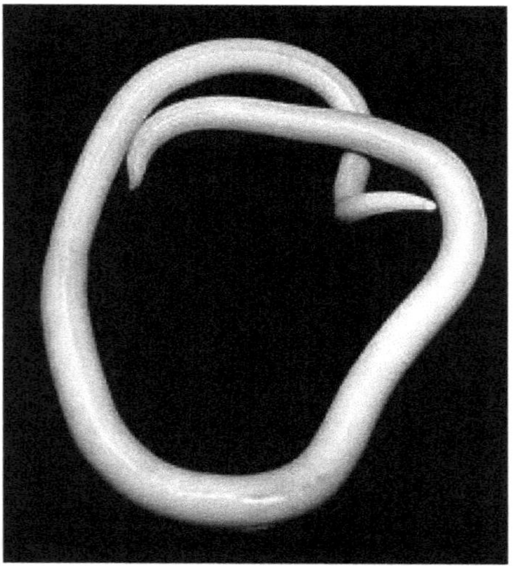

FIGURE 39. **Female *Ascaris lumbricoides*.** The most widespread human nematode parasite. Note the lack of segmentation. As with many nematodes, their lack of physically defining features illustrates the difficulties involved with nematode taxonomy.

---

[10] Bar coding is a means of recognizing taxonomic relationships among groups of organisms from patterns of DNA sequences. Similarity of nucleic acid sequencing is used as a measure of conspecificity.

The mammalian kidney parasite *Dioctophyme renale* is another large nematode species that is found in a variety of wild and domestic mammals. Like *Ascaris* the adult female kidney worm is larger than the male and may grow up to 100cm (40") in length and 6mm in diameter. However, the record for nematode size is held by the giant nematode *Placentonema gigantissima*, a parasite of blue whales which may reach 8.4 metres in length and 2.5 cm in diameter. As its name suggests, it is found in the placenta of sperm whales but has been discovered in many parts of the sperm whales' reproductive system.

Aside from massive differences in adult size between many free-living and parasitic nematode worms, a feature of the phylum Nematoda is that they often lack obvious physical characteristics that can be used to distinguish among different nematode species. A glance at the picture of the parasite *Ascaris* (FIGURE 39) illustrates this point, revealing little more than a feature-less threadworm lacking segmentation and other physical characteristics that usually separate one species from another. Obviously adult size is an important defining feature in their taxonomy. However, in a 'field situation' where the age of specimens may be unknown, it is easy to see why genetic configuration rather than morphology is often used to differentiate among nematode groups or individuals due to their similar appearance. The downside of the genetic approach is the resulting multiplication of species numbers that might jump from twenty-plus thousand to as much as a million. This issue is the focus of a fairly recent paper by Michael Hodda (Hodda 2022) that addresses the problems associated with the taxonomy of such a massive group of animals that so often resemble each other.

Nevertheless, the economic importance of the phylum Nematoda continues to drive the need to understand the taxonomy of this ubiquitous yet elusive group of organisms.

For example, recent estimates of agricultural crop losses due to nematodes exceed $300 billion annually, underlining the need for the correct taxonomic identification of nematodes so that the appropriate treatment for their control can be employed. The same applies to the several thousand nematode species that are parasitic on humans as well as their pets and livestock. Again, correct species recognition is critical for the threat to be identified so that effective remedial measures can be deployed.

In several respects the members of the phylum Nematoda are emblematic of a secretive creepy crawly invertebrate world on the planet Earth; a parallel universe shared with humans. Nematodes seem to be everywhere on the planet despite the fact that we are so rarely aware of their presence. Indeed, they are so ubiquitous that it was suggested by U.S. Department of Agriculture scientist Nathan Cobb in 1914 that if the land and water masses of the world were to disappear, leaving only nematodes, these residual worms would still leave an outline of all the geographic features such as plains, mountains and oceans that constitute our planet!

The ubiquity of nematodes on Earth signifies their remarkable tolerance of extreme environments and climatic conditions. This includes an apparent capability to shut down their metabolism indefinitely under adverse conditions. This ability to enter a state of suspended animation has been spectacularly demonstrated in the case of the nematode *Panagrolaimus kolymaensis* which has been 'revived' following its retrieval from Siberian late Pleistocene permafrost estimated to be more than 45,000 years old!

## Worms in Space?

Considering the ability to tolerate extreme environmental conditions and to suspend their metabolism indefinitely, you could be tempted to believe that nematodes might have the capacity to survive on other planets. If that thought has ever crossed your mind you are not alone.

In 1963, South African scientist Sydney Brenner first proposed the use of the nematode *Caenorhabtitis elegans* as a prototype for testing the effect of environmental stress and other factors on developmental biology, neural physiology and genetics. This nematode species has the distinction of being the first organism to have its whole genome sequenced as well as serving as a prototype for mapping neural networks. This broad spectrum of work earned Brenner the 2002 Nobel Prize in physiology as well attracting the attention of the space agencies operating the International Space Station. Batches of *C. elegans* became frequent visitors to the International Space Station. Among topics of investigation were the more immediate issues such as determining the effect of the physical stresses of space flight on biological systems. These included the effect of powerful G (gravitational) forces associated with rocket thrust as well as weightlessness once escape from earth's gravitation had been achieved. An inadvertent test of temperature and other extremes occurred with the catastrophic explosion and disintegration of the space shuttle "*Columbia*" in 2003 with the tragic loss of its seven crew members. The only living survivors of this disaster were the *C. elegans* nematodes that accompanied the astronauts on their mission.

A more long-term goal behind the use of nematodes to assess 'space viability' relates to the exploration of what many of us would see as a remote objective: the colonization of another planet such as Mars. Clearly some explanation is needed here. In addition to their extraordinary capacity for survival under adverse environmental circumstances, another characteristic of the earthly existence of nematodes has been their ability to act in concert with numerous and varied life forms in an unparalleled range of physical environments. The world-wide interaction of nematodes with other biota has played roles including the cycling of important nutrients such as atmospheric nitrogen and the decomposition and turnover of a broad range of other biota.

While some would say that it is too early to be thinking about the establishment of life on other planets, there is certainly a case to be made for including nematodes as potential pioneer organisms in such an initiative. The word 'pioneer' can be seen as important in this context. The remarkable ability of nematodes to survive and even thrive in seemingly unforgiving habitats could help establish conditions and resources that would increase the chances of survival of a broader flora and fauna.

# Phylum Mollusca

Over 85,000 living members of this phylum are currently recognised, making it the second-largest invertebrate phylum after the Arthropoda. Although many of their number possess protective shells, they are primarily soft-bodied animals, which accounts for the phylum name Mollusca derived from the Latin Molluscus meaning 'soft'. The Mollusca phylum is subdivided into the following seven classes:

Aplacophora are worm-like shell-less animals that occupy a deepwater benthic habitat in ocean waters. They are thought to derive from ancient Silurian ancestry over 400 million years old.

Monoplacophorans have a single cap-like shell and were believed to be extinct until the discovery of the 'living fossil' *Neopilina galathaea* in 1952.

Polyplacophoa (chitons) are intertidal forms that may grow up to 20cm (8") in length possess a dorsal shell consisting of eight plates.

Bivalvia. Marine and freshwater with calcium carbonate shells consisting of two hinged parts. They include mussels, oysters, clams and scallops. These bivalves support a 440 million dollar cultivation industry

Gastropoda. Snails, slugs, conches and sea butterflies. They include the only terrestrial molluscs (slugs and snails).

Scaphopoda. Characterised by a tusk-shaped shell that earns these molluscs the common name 'tusk shell' or 'tooth shell'. The foot of scaphopods is modified into tentacles for prey capture.

Cephalopoda. Squids, octopuses. cuttlefish, nautilus. These marine predators include the largest of all members of the phylum Mollusca. The female giant squid *Architeuthis dux* may grow up to 12m (40') in length, including both mantle and tentacles.

## A Deadly Octopus.

Although the blue-ringed octopus *Hapalochlaena lunulata* (FIGURE 40) is at the lower end of the octopus size range (12-20cm in length) it carries a deadly venom that can kill a human following total paralysis. Death may occur within 15 minutes of attack. About 12 human deaths have been attributed to this species. The primary active compounds of *H. lunulata* venom are tetrodotoxin tryptamine and dopamine.

FIGURE 40. **Blue-Ringed Octopus** *Hapalochlaena lunulate.* A small but deadly cephalopod found in the Asian Pacific.

# General Reading.

R. Howells (2022). The Language of Bees. Publ. Parthian Books, Cardiff. ISBN 978-1-913640-69-9. 78pp.

T.D. Seeley (2010). Honeybee Democracy. Princeton Univ. Press. Princeton New Jersey, U.S.A. ISBN 978-0-691-14721-5. 280pp.

T.D. Seeley. (2025). Honeybee Ecology: A Study of Adaptation in Social Life. Princeton Univ. Press. Princeton New Jersey, U.S.A. ISBN 978-0-691-2736-5. 280pp.

# References.

C. Bettin, J. Oehlmann, E. Stroben, (1996). TBT-induced imposex in marine neogastropods is mediated by an increasing androgen level. *Helgoländer Meeresuntersuchungen.* **50** (3): pp 299–317.

C.A. Callahan, C.A. Menzie, D. E. Burmaster, D.C. Wilborn and T. Ernst (1991). On-site methods for assessing chemical impact on the soil environment using earthworms: A case study at the Baird and McGuire Superfund Site, Holbrook, Massachusetts. Environ. Toxicol. Chem. **10**(6). pp 817-826. https://doi.org/10.1002/etc.5620100612.

M. Hodda (2022). Phylum Nematoda: trends in species descriptions, the documentation of diversity, systematics, and the species concept. Zootaxa. doi: 10.11646/zootaxa.5114.1.2. pp. 290-317.

M. Kaplan (2013). First venomous crustacean discovered. Nature. https:// doi.org/10.1038/nature 2013.13985.

T.M. Lilley, L. Ruokolainen, A. Pikkarainen, V.N. Laine, J. Kilpimaa, M.J. Rantala and M. Nikinmaa. (2012). Impact of tributyltin on immune response and life history traits of *Chironomus riparius*: single and multigeneration effects and recovery from pollution. Environ Sci Technol. **46**(13), pp 7382-9.

P.E. Marek, B.A. Buzato, W.A. Shear, J.C. Means, M.S. Harvey and J. Rodrigues (2021). The first true millipede – 1,306 legs long. Scientific Reports, **11**(1). Article number 23126. https:// doi.org/10.1088/s41598-021-02447-0.

M. Rutgers, A.J. Schouten, J. Bloem, N. van Eekeren, R.G.M. de Goede, G.A.J.M. Jagersop Akkerhuis, A. van der Wal, C. Mulder, L. Brussaard and A.M. Breure. (2009) Biological measurements in a nationwide soil monitoring network. Eur J Soil Sci **60**:820–832.

P. Schultheiss, S.S. Nooten, R.Wang, M.K.L. Wong, F. Brassard and B. Guénard (2022). The abundance, biomass, and distribution of ants on earth. Proc. Natl. Acad. Sci USA. **119**(40): e2201550119. doi:10.1073/pnas.2201550119.

S. Tischer (2009). Earthworms (Lumbricidae) as bioindicators: the relationship between in-soil and in-tissue heavy metal content. Pol J. Ecol. **57**: 531-541.

C. Vogt, C. Nowak, J.B. Diogo, M. Oetken, K. Schwenk and J. Oehlmann. (2007). Multi-generational studies with *Chironomous riparius* – Effects of low tributyltin concentrations on life history parameters and genetic diversity. Chemosphere **62** (11), pp 2192-2200.

B.E. von Reumont, A. Blanke, S. Richter, F. Alvarez, C. Bleidorn and R.A. Jenner (2013). The First Venomous Crustacean Revealed by Transcriptomics and Functional Morphology: Remipede Venom Glands Express a Unique Toxin Cocktail Dominated by Enzymes and a Neurotoxin. Mol. Biol. Evol. **31**(1): 48-58. doi: 10.1093/molbev/mst 199.

D.A. Wright (1975). Sodium regulation in the larvae of *Chironomous dorsalis* (Meig.) and *Camptochironomous tentans* (Fabr.): The effect of salt depletion and some observations on temperature changes. J. Exp. Biol. **62**, pp 121-139.

D.A. Wright and W.H. Roosenburg (1982). Trimethyltin toxicity to larval *Uca pugilator*: Effects of temperature and salinity. Archiv. Environ. Contam. Toxicol. **11**, 491-495.

Other books in this series by David Wright published by Cambria Publishing.

ISBN: 978-1-0685688-7-9

Pages: 222

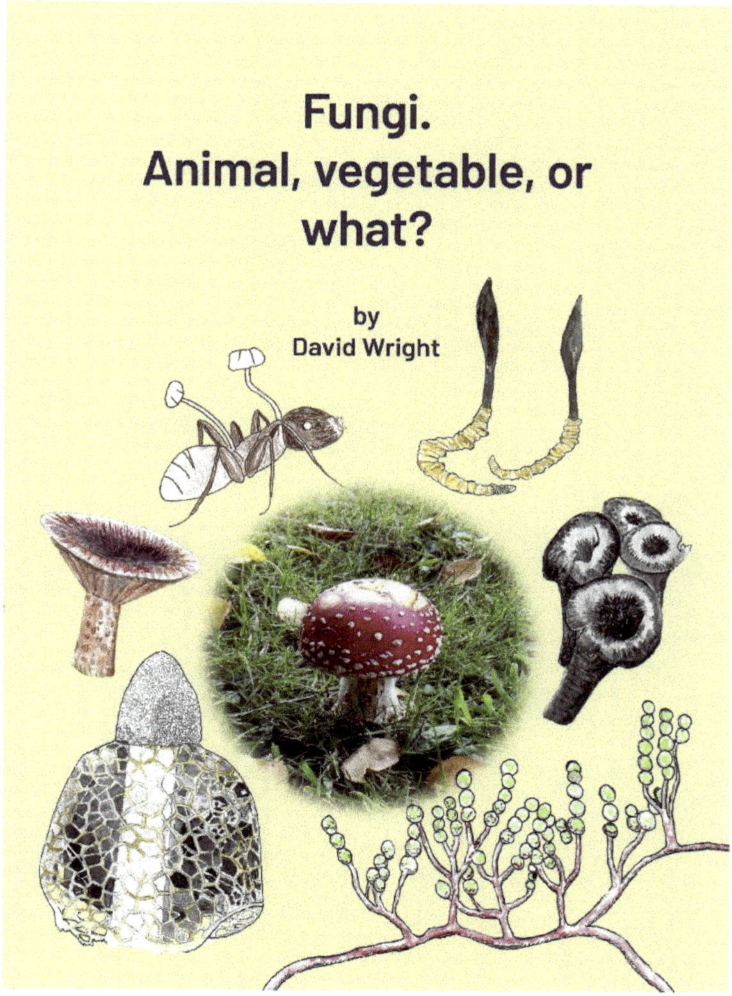


ISBN: 978-1-9192698-9-4

Pages: 174